Tough Words
for American Industry

Tough Words for American Industry

Hajime Karatsu

Foreword by
Norman Bodek, President
Productivity, Inc.

Productivity Press
Cambridge, MA
Norwalk, CT

Productivity Press
P.O. Box 3007
Cambridge, MA 02140
(617) 497-5146
or
Productivity, Inc.
101 Merritt 7 Corporate Park
Norwalk, CT 06851
(203) 846-3777

Library of Congress Catalog Card Number: 87-43206
ISBN: 0-915299-25-9

Cover design: Bill Stanton
Typeset by Rudra Press, Cambridge, Massachusetts
Printed and bound by The Maple-Vail Book Manufacturing Group
Printed in the United States of America

Library of Congress Cataloging-in Publication Data

Karatsu, Hajime, 1919-
   Tough Words for American Industry.

Translation of: Kūdōkasuru Amerika sangyō e no chokugen.
   1. Japan—Foreign economic relations—United States.
   2. United States—Foreign economic relations—Japan.
   3. Industrial productivity—Japan.   4. Industrial productivity—United States.
   5. Quality of products—Japan.   6. Quality of products—United States.
   I. Title
HF1602.15.U6K3613   1988   338.0973   87-43206
ISBN 0-915299-25-9

88   89   90   10   9   8   7   6   5   4   3   2   1

Contents

Table of Figures

Table of Photographs

Publisher's Foreword

You're probably not going to like what you read in this book. The author, a professor and consultant well-known in Japan, may often seem arrogant, and a description of America's problems may not be what you wanted to hear. So why are we bringing this book to America?

Because Karatsu's criticism is absolutely valid. It may be just as unpleasant as castor oil, but the message here is a remedy we need to remove the scales from our eyes. It shocks us to view the depth of our industrial decline, but it's a nasty fact we have to recognize before we can start doing something about. And, like it or not, the Japanese do have a thing or two to teach us.

The fallacy of short-term thinking is one such lesson Karatsu drives home — a timely message in light of the recent volatility of the stock market. The profit motive has led a number of major American corporations to at least partially abandon domestic production in favor of offshore plants with lower labor costs. The result: empty American factories, idle workers, and cash and technology handed to developing nations that may someday dominate us in the world market when we no longer know how to make things at home.

The Japanese approach things differently. Corporate executives everywhere like a healthy profit. But Japanese managers are more interested in what they need to build a base for the stability and well-being of their company for many years in the future. It may cost something on the front side to strictly monitor product quality and to develop the skills and abilities of employees at home. It may take time and money to build a structure in which employees at all levels truly participate in corporate and personal improvement through reward-based suggestion programs and QC circles. But these investments pay off — no one can deny the quantum leap made by Japanese industry in the decades since World War II.

America's corporate managers deserve this criticism for their shortsightedness. The CEOs who have allowed our industrial base to disintegrate at home need a Karatsu to chew them out for their lack of vision.

The Japanese are not spotless angels, either. They *have* restricted foreign competition, making it difficult for the U.S. and other countries to sell in Japan. It is commonplace for book distributors, for example, to sell books at the old higher yen-to-dollar rates, thereby forcing prices above competitive levels. And a Japanese electronics firm will extract a royalty from an American company wishing to use one of its designs, even though it might make the design available to another Japanese company at no charge.

A number of American companies *are* doing well in Japan. Those are the ones that have made the effort to discover and supply the products and the quality the Japanese are looking for — using the "market-in" approach that has made Japanese products so successful with American consumers. But there are still too many who seem to expect the Japanese to come to us. We're upset when a nation that drives on the left doesn't want our cars, built for driving on the right. We're offended that the Japanese don't speak English to us when we come to sell them something!

Karatsu's message is clear: Americans should think twice before blaming the Japanese for our faltering economy and trade deficit. He brilliantly pinpoints the problems of American manufacturing and the accuracy of his assessment is an unfortunate reflection of our own arrogance — an attitude that cannot continue if we are to regain our position as a dominant industrial nation. We may not particularly like this medicine, but we need more than sugar pills to make us healthy again. Karatsu wants to make us mad with these tough words — to shake us out of the kind of complacency that has led to the downfall of the great civilizations of the past. If they provide the necessary kick in the pants, we should be grateful to him.

I would like to thank Connie Dyer and Karen Jones for the editorial work on *Tough Words*; Esmé McTighe and Patty Slote for production; Caroline Kutil, Michele Seery, and Susan Cobb of Rudra Press for typesetting and layout; and Bill Stanton for his book and cover design.

Norman Bodek
President, Productivity, Inc.

Preface to the Japanese Edition

T he starting point of any economy is manufacturing — the making of things. Money not backed up by manufacturing is nothing more than wastepaper. To forget this very basic truth in the conduct of foreign relations or politics is to court economic destruction. In today's world this possibility is a more pressing problem than the threat of nuclear war; it directly affects the lives of people everywhere. The world economy is headed at an ever-increasing speed into the jaws of crisis. One of the most telling signs of this phenomenon is the trade friction that now exists between Japan and the United States.

In September 1985, the finance ministers of the Group of Five industrial nations took steps to weaken the dollar on the international currency exchange market. This action was designed to effect a 30 percent cost reduction in the products of U.S. industries. Since that time, however, I have not heard any tales of a surge in sales of American products in international markets. The reason is clear: American manufacturing industries have been sacrificed to the "money game" and have become hollow, leaving large gaps in the economy. There is nothing left to sell. Under these circumstances, there is no possibility that the trade imbalance, which stands at the heart of the problems between the United States and Japan, can be eliminated — no matter how high the yen is driven up.

The forced appreciation of the yen that was induced in[?] was a major economic experiment on the part of econom[?] around the world. This experiment is still underway and has in[?] flicted a good deal of pain on the Japanese industrial world. For this reason, we should feel obligated to correctly analyze the data from the experiment, and to determine what it tells us about the directions we should take in the future. To do anything less would simply cause more unnecessary pain to those who are being sacrificed to pay for the experiment.

In spite of this obvious necessity, recent international arguments have centered on how to cut Japan's trade surplus — and not to determine what we can learn from the current abnormal economic situation. Only by removing cancerous cells can the malignancy be kept from regenerating. Any attempt to treat this economic cancer that loses sight of the fundamental starting point, manufacturing activities, will not really get at the root of the problem. Proposed solutions that do not deal with the essence of the problem will only result in spreading the cancer. The stimulation of internal demand in Japan falls into this category. Even if we were successful in expanding domestic demand, we would be kidding ourselves if we believed that this alone would result in a new influx of American products in the Japanese market.

This book is a collection of my essays published over the past few years in such magazines as *Voice*. I have reworked them into their present form and tied them to my own opinions and views concerning the directions we should take.

The primary focus of the book is on U.S.-Japanese economic relations, but it addresses larger issues. My goal has been to serve as a sort of devil's advocate, by challenging everyone in the manufacturing world to reconsider their opinions on the starting points that will best bring about economic health in the future.

All economics boils down to a struggle between theory and reality. This struggle does not occur on a theoretical front. If reality does not turn out the way the system says it will, the system itself is nothing more than a dangerous form of empty theorizing. This is true no matter how splendid the underlying logic. The act of making things includes a number of factors that can be classified as neither black nor white but gray. In this book I stress the need for a methodology that will teach us how to deal with these. This is economic reality and is what has brought Japan to success. The basis for economic growth is manufacturing and it is the principle of productivity that will enrich everyone. My main wish is that Americans, and the other people of the world, will rethink these crucial questions.

To help reach this goal, I have pulled no punches in this book. Doubtless there will be some who feel that my approach is rude, or even arrogant. But that is my intention; I have merely tried to report reality just as it is. Emotionalism is dangerous and unprofitable in this struggle.

Since World War II Japan has worked as a country possessed to close the wide technological gap that separated its industries from those of the West. We have finally made it, and can now take our place among the world's top industrial nations. This book should be taken as the frank opinions of one of those involved in this struggle.

Hajime Karatsu

Tough Words
for American Industry

CHAPTER 1

The Hidden Patterns Behind U.S.-Japan Trade Friction

In November 1983, when President Ronald Reagan was visiting Japan, I wrote the following words in the *Japan Times*:

> In my opinion, the United States should, if anything, increase its imports from Japan. If we total the import/export balances for the two countries, we will certainly find that Japan has an export trade surplus — but, despite that surplus, the average Japanese person spends $204 per year on products imported from the United States, while the average American spends no more than $155 on products imported from Japan. Couldn't we truly say, then, that it is the Japanese who are buying too much?

American journalists accompanying the president must have been taken aback somewhat by such a forthright opinion from a Japanese commentator. Yet there should be nothing surprising about the conclusions I drew. In fact, they are natural enough, when one considers that the population of Japan is but one-half that of the United States. If absolute parity were the major goal of trade relations between the United States and Japan, attaining it would require that the Japanese spend only half as much as the Americans on a per capita basis. It is totally unreasonable to expect this to happen and I said so in that essay.

In recent months, articles on the trade friction between Japan and the United States have been more common than the cawing of crows. I, for one, am getting sick of them. The situation gets really complicated when one considers other problems, such as the rights of American attorneys to practice in Japan or copyright protections for computer software.

Even so, however, a careful analysis of these problems leads one to feel that a good deal of what is being said qualifies as false charges. Many of these are made by politicians playing to their respective galleries.

Among the most common of these charges is that Japan has closed its domestic market to U.S. producers. Sales of beef and oranges are inevitably brought up here as an example. The facts are otherwise. As shown in Figure 1, 1985 statistics reveal that some 78 percent of the beef exported by the United States has been going to Japan. Japan's share of the orange export market has been *as high as 26 percent.* Therefore, Japan has been America's best customer for both beef and oranges.

Furthermore, let us not forget that the price of American beef is high in Japan. Ironically, the total liberation of the Japanese market would result in the triumph of lower-priced Australian and Argentinean beef at the expense of American beef. The United States itself maintains import restrictions aimed at beef from these two countries for this very reason.

A sober look at the actual figures clearly shows — at a level a child could understand — that these problems are political by nature. Even though they hear the facts again and again, many Americans continue to arrive at incorrect conclusions. This ignorance extends throughout the U.S. body politic.

During a recent stay in Washington, D.C., I learned something that truly distressed me. Members of the staff at the Brookings Institute sent a trade questionnaire to members of the U.S. Senate and House of Representatives. One of the questions dealt with Japan's rank as a consumer of American beef and asked whether Japan was the *best* customer, an *average* customer, or a *low-ranking* customer. Seventy percent of the respondents are said to have circled "low-ranking," and not even 10 percent of them answered the question correctly.

A lie, if repeated often enough, will eventually be taken for the truth. Even in Japan, most people are surprised to hear the figures quoted above concerning beef, which shows that the situation has reached crisis proportions.

	Dollar Amounts	Volume
Soybeans	25.5%	25.9%
Tobacco*	19.9	18.5
Beef**	77.3	78.3
Pork**	53.2	26.5
Oranges***	30.2	26.3
Lemons***	81.9	76.9
Grapefruit***	56.5	52.9

* : Before processing
** : Fresh and frozen
*** : Fresh
Source: U.S. Department of Agriculture

Figure 1. Japanese Shares of U.S. Agricultural Exports (1985)

	Amount Exported (billions of dollars)	Percentage
Japan	5.4	18.6
Soviet Union	1.9	6.6
Holland	1.9	6.4
Canada	1.6	5.6
Mexico	1.4	5.0
Totals	**29.0**	**100**

Source: JETRO, *Quick Reference Handbook of Economic Relations Between the United States and Japan*

Figure 2. Major Importers of U.S. Agricultural Goods (1985)

One scene to which we have been frequently treated on Japanese television depicts a physical attack on a Japanese automobile. In a depressed part of the Rust Belt, men — apparently out of work — are shown destroying a Japanese car. What better way to suggest that Japan must bear the brunt of responsibility for the rise in the unemployment rate in the United States! As revealed here, the workings of public opinion are frightening indeed. Unfortunately, the Japanese mass media that show these film clips tend to perpetuate the transmission of this type of warped information.

What is the reality here? Do Japanese exports really have the power to wreak chaos on the American economy? Look carefully at the following calculation before you automatically answer "yes."

Foreign imports comprise, by and large, 9 percent of U.S. GNP. Japan's share amounts to about 10 percent of the whole. Therefore, Japanese exports come to no more than 0.9 percent, at the very highest, of overall U.S. GNP. This figure, about 1 percent of the entire nation's GNP, must be considered negligible. Any claim that it can alter the course of the entire U.S. economy can only be termed "exaggerated."

These principles also apply to specific aspects of the U.S. economy. Three of these are particularly illuminating: automobiles, televisions, and the defense industry. An examination of recent trends shows the futility of shaping political responses to economic problems.

A careful look at Leontief's input/output tables supports the conclusion that the increase in exports of automobiles from Japan to the United States has not had a significant impact on American industry. At best, the effect has been marginal.

I've no doubt that Leontief's tables are part of the curriculum at virtually all Japanese universities. It seems to me that we should at least occasionally see an economic simulation of U.S.-Japan trade based on them; I find it somewhat distressing that there has been absolutely no evidence of this. If this were a defense problem, we would see a real outpouring of debate, but the trade issue appears to be another matter altogether.

Import Restrictions That Backfire

Voluntary export restraints and import restrictions on the sale of Japanese automobiles have not had their intended effect. It is a common belief in the United States that the institution of voluntary restraints on Japanese car exports played a key role in accelerating inflation in that country in the early 1980s.

At that time, buying a Japanese car in the United States involved the paying of a premium above the list price. Even economy cars such as the Nissan Sunny or the Toyota Corolla, priced at around $7,000, required a payment of from $1,500 to $2,000, or from ¥300,000 to ¥500,000, above the manufacturer's recommended price. With these bonuses, export restrictions had U.S. dealers of Japanese cars laughing all the way to the bank.

The premium pricing of Japanese cars might have been expected, but then American car manufacturers started to get into the act, raising their own prices to earn themselves similar premiums. This had a considerable ripple effect, because the American automotive industry influences the direction of the country's entire economy. The price run-up for Japanese cars may have had no real effect on the American economy, but when American automakers also raised prices, they contributed mightily to the development of inflation in the United States. When the stronger yen forces up prices of Japanese cars once again, prices of American cars will also increase.

It was the American consumer who suffered most from these export restraints. U.S. automakers made out like bandits. Furthermore, these manipulations went virtually unreported in Japan. As a result the trusting Japanese were made to play the part of the fool. It's no wonder that Americans hold us in contempt.

Slap on import restrictions and make a lot of money! Naturally, this particular turn of events restored American economists to a state of glowing health. To miss this type of golden opportunity would have cast their own skills as managers into serious doubt.

And even if the intentions were not all bad, there is no doubt that the experience has helped foster a denial of the principles of free competition in America's subconscious. The rise in protectionism, the passage of local content laws, and the erosion of antitrust laws are political manifestations of this trend. These are, of course, some of the dangers in an economy based on the

Japanese cars are conspicuous at new American housing developments

ideas of liberalism. If current trends are allowed to continue, the structure of U.S. industry will evolve into a prime example of the controlled economy. And this will inevitably bring great stress to the economy in the future.

Opposition to protectionism *has* appeared in the United States. It stems from the instinctive panic that has arisen among those involved in distribution and is not limited to the so-called intelligentsia. This all came to the surface as an issue in the presidential election of 1984. It was embodied in what might be called a grassroots antiprotectionist movement and was reflected in the positions of several candidates for the Democratic presidential nomination. Gary Hart was a prime example. We should pay careful attention to this. The move-ment opposes the local content laws, and holds instead that the best way to revitalize American industry will be to provide the least restrictive framework for its operation.

There is considerable danger in the idea that playing around with politics will lead to profits. The United States has been unabating in its criticisms of Japan, but anyone who stops to consider can see the hazard involved in continuing down that path. Many people in the United States have begun to realize this fact.

Unfortunately, American corporate management continues to act in a selfish manner that sidesteps the realities of productivity issues. The American color television market shows this well. Competition now occurs primarily between televisions with American brand names — made abroad — and those with Japanese brand names, made in the United States.

Japanese television manufacturers, stung by anti-dumping laws and a variety of other regulations, have set up production facilities in the United States. American manufacturers, though, have gradually shifted their production facilities abroad, in search of lower labor costs.

This may not appear especially strange to Americans, but from a Japanese perspective it is truly an indication of American selfishness. Above all it shows that Americans believe that everyone in the world is obligated to play the game by the American rules. How long, I wonder, do they believe this kind of thing can go on?

The same thing has now started to happen with automobiles. Virtually all of the Japanese automakers have started manufacturing in the United States. But the American automobile companies have their eyes on the cheaper labor costs available in Mexico and have begun to shift their production offshore. I predict that before long we'll have the same situation that exists with color televisions — market competition between Japanese cars made in the United States and American cars manufactured abroad.

The defense industry is the best example the excessively self-centered philosophy of American corporate management. It is in a real uproar. The furor has been caused by the fact that the Americans have suddenly noticed that some 70 percent of American semiconductors are now produced abroad. This is a serious problem, and one that has a direct bearing on the levels and stability of U.S. defenses in these days of conflict between the United States and the Soviet Union. It is a fundamental problem and not one for which Americans can find an easy scapegoat. It touches at the very heart of the concepts held by American corporate managers.

Political reactions are one thing, but what do the statistics say about the other side of the coin — Japanese investment in the United States? If we investigate a little more closely, we will find that figures reveal impressive activity on the part of Japanese firms working in the United States.

According to 1983 data from the U.S. Department of Commerce published by the Japan External Trade Organization (JETRO), the total value of exported goods produced by foreign firms working in the United States was approximately $53 billion (see Figure 4). Of that amount the largest share — $22 billion — was racked up by Japanese companies. It represents some 11 percent of the entire value of American exports. This figure is, of course, growing; it was predicted to exceed $24 billion in 1986. This indicates the extent of the efforts that Japanese firms have made to improve the account balances kept in U.S. trade books.

American factories planned by the Japanese automakers will begin production by the 1990s and will more than double the share of American exports held by Japanese plants. Similar investments will occur in other industries. If figures such as these are considered individually for each industrial sector, one cannot help but wonder exactly what it is that has created U.S.-Japanese trade friction in the past few years.

As far as I'm concerned, the truth is that Japan has been paying for the poverty of ideas of American management. The next section will give you another example of what I am talking about.

Offshore Plants and Manufacturing Ignorance

Americans refer to overseas production as "offshore manufacturing." A 1985 Japanese television documentary portrayed the possibilities involved in such manufacturing.

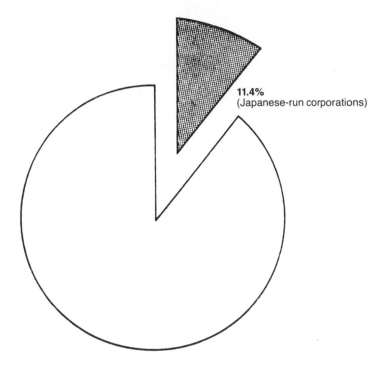

11.4%
(Japanese-run corporations)

Source: JETRO, *Quick Reference Handbook of Economic Relations Between the United States and Japan*

Figure 3. Portion of Total U.S. Exports Received by Japanese-Run Corporations (1983)

	Amount Exported (billions of dollars)	Percentage
Japanese firms	22.9	42.5
French firms	9.2	17.1
Canadian firms	4.2	8.8
British firms	3.3	6.1
German firms	2.7	5.0

Source: JETRO, *Quick Reference Handbook of Economic Relations Between the United States and Japan*

Figure 4. U.S. Exports to Foreign Firms by Country (1983)

In the story, an American semiconductor manufacturer undertook to make semiconductor chips abroad. First, it contracted its design work in Israel. The design would be checked by the main office in Silicon Valley, and once approved, masks would be made for fabrication of the silicon chips. The chips would then be transported by air to a factory in Southeast Asia, where the labor-intensive assembly processes would be carried out. The completed products would once again be shipped by air to the main factory in the United States, where a variety of tests would be conducted. Those products that passed the inspections would then yet again be distributed by air to markets throughout the world. The picture evoked was truly that of leading edge technology that knew no geographical boundaries.

Those who heard this tale would believe, doubtless, that this was a case of a lightweight and high value-added product manufactured in such a way that each step of production could be carried out in that part of the world where conditions were the best for the individual process. It must have thus seemed to be the most rational system possible. The television show itself contributed to this impression, by giving an impressive-looking introduction to state-of-the-art technology that uses the entire world as a stage.

The whole thing sounded like something thought up by a bright business student intent on influencing the course of manufacturing theory. But my opinion was somewhat different: I thought it went a bit too far to propose such an approach without an inkling of what goes on in a manufacturing plant. Is it really possible to efficiently control the manufacturing processes using such a system?

Let us look at the Japanese experience. The manufacturing is the best in the world, but this quality has only been achieved as a result of total quality control (TQC) programs that encompass everyone involved in the manufacturing process; such programs are, in short, an integral part of the manufacturing control system. It is a major mistake to think that every plan hatched by someone behind a desk will work out the way it is supposed to. Plans such as that in the program are, furthermore, subject to incredible snafus from even minor foul-ups, and it can take significant expenditures of both time and work to get them back on track once they have derailed.

Here is one horrible example. Several years ago defects suddenly began popping up regularly in the products of an American firm that had been selling quite successfully in Japan. It took two months for the problems to be ironed out. The people who suffered most keenly from these problems, of course, were the Japanese who had bought the products. They were notified through a single, simple letter only that their supply would be cut off for the duration of the problem. A friend of mine has that letter and has kept it as proof of the American proclivity towards one-way communications. No one will continue to buy something if a continuous supply of that product cannot be guaranteed. The average cost of a semiconductor chip is about two dollars. But if lack of a chip has made it impossible to make delivery on several pieces of equipment worth several hundred thousand dollars, you can be certain that the person who was stung will remember his anger.

Certainly there are many American products that are well conceived and have outstanding functions. But no one will buy them until they have been assured that they can get the number they need, when they need them, and that the products will not break down. Of course, there are American companies that do not differ from Japanese firms in this respect, but I speak from experience when I say that it is true that Japanese and American firms often differ in their response to product failures.

The Toyota production system has become famous abroad as well as in Japan. But even the Toyota system cannot be considered complete until the number of defective parts drops to zero. The kanban production control method used at Toyota would not amount to anything if parts ordered through it failed to function properly once they had been assembled. Through the use of local content laws the United States is attempting to force Japanese companies to use American parts; such solutions are being advanced by people who have no knowledge of what goes on at a manufacturing plant. Manufacturing is a struggle that takes place at the plant; it is not a matter of juggling figures at a desk.

Failure to recognize this principle also extends to members of the American media. At a conference held in 1983, Japan claimed world leadership in the semiconductor industry in developing the first DRAM (dynamic random access memory) chip with a capacity of one megabit (1 M). The size of semiconductor memory devices has steadily increased. They were first marketed with capacities of 64 kilobits, and then 256 K. The development of a 1 M chip had been eagerly awaited.

One newspaper, however, carried the news of this development under the headline, "Still Another Cause for Friction?" That was a totally wrongheaded approach. This piece of news should have been treated much more positively; here we had a case where Japan, often criticized for imitating the technology of other countries, had come out with a breakthrough. The headline should have reflected well-deserved pride. The attitude of the newspaper — that the bigger the fire that takes place somewhere else, the more interesting it becomes — really put a damper on what should have been positive news concerning Japanese accomplishments.

At that time, I rewrote the essay quoted from the beginning of this chapter and the more detailed version was published in the Asian edition of the *Wall Street Journal*. I received positive responses to that article from American readers in Washington, who said that I had done a good job of presenting Japanese opinion on these issues.

Media silence on Japanese technological developments is also a problem. After a thorough preliminary study, a Japanese research group in 1982 launched a major project aimed at developing a "fifth generation" computer using artificial intelligence. This provided major stimulation for similar research projects throughout the world. The governments of several other countries, including Britain, France, and the United States, were spurred on by this announcement to start their own projects.

The Japanese development group, called the Institute for New Generation Computer Technology (ICOT), is housed in the International Building in Tokyo's Mita district. It is surprising how little one reads about the openness this organization has shown toward researchers from overseas. Visitors from abroad are always welcomed and invited to stay at Japanese expense for periods of approximately one month to encourage the free flow of discussion and debate. Each year ICOT invites a number of foreign researchers at its own expense; in fact, there were more than a thousand applicants for the international symposium it held in 1984. Information concerning this project has been openly available to foreigners.

The free flow of research information may be going one way, however. In recent years I have started to hear scattered tales about how Americans have kept Japanese scholars from participating in academic conferences. Doubtless this is linked to trade friction. It is said in Japan that a person who hates a Buddhist priest will hate everything about him — even his hood. It smacks of a children's quarrel, though, to try to keep Japanese scholars out of the world's intellectual community because of a trade squabble with Japan. The source of this quarrel, in my mind at least, is strictly one-sided and stems from politicians who do nothing but dance around the altar of short-term profits. One wonders what future historians will have to say about the situation we are faced with today.

Beef, Oranges, and Much More

Japan is importing more than beef and oranges. In fact, Japan is the best customer that American farmers have, as Figure 1 shows. Some 18 percent of American agricultural exports, more than three times those sent to the Soviet Union, go to Japan. It would seem a matter of common sense that such a good customer should be entitled to a Christmas present every now and then. But what has actually happened lately is quite the opposite — at every opportunity America has demanded gifts from Japan. Some might ascribe this attitude to upcoming elections in the United States, but even considering this fact, American demands seem a little strident.

As an exporter of agricultural products and raw materials, and an importer of Japanese manufactured goods, America would appear to be nothing more than a developing nation. Such an impression may offend some people, so I will hasten to make a little-known point: Did you know that the largest customer in the world for U.S. manufactured goods is a firm in Japan? The customer is Japan Air Lines. In 1986, JAL already owned 68 jumbo jets (Boeing 747s) and was ordering more. With the exception of U.S. airlines, this is the largest fleet of any privately owned airline in any country in the world.

So there is no truth to the idea that the only things Japan buys from the United States are agricultural products. Figure 5 illustrates this. A total of 52.3 percent of all U.S. exports to Japan are manufactured goods. U.S. manufactured goods have lost a good deal of their competitive edge in the international market and American exports to virtually all countries have fallen off. In the midst of this general decline, exports to Japan alone have continued to grow.

Figure 6 shows total American exports broken down by destination. Take a close look at it. Japan ranks second, right after Canada. This is perhaps as it should be, since Canada occupies the same land mass as the United States and is a next-door neighbor. But Japan is second; the dollar value of its imports from the United States nearly equals those of Mexico (number 3) and West Germany (number 5) combined. Figure 7 graphically depicts Japan at the head of the growth trend in U.S. exports over much of the last decade.

Very few members of Congress are familiar with these figures. And 99 out of 100 people with an interest in the trade problem are surprised to hear this information. It is scarcely known in Japan, a fact that bothers me a great deal.

When addressing this lack of knowledge, I use the familiar lesson from Plato's allegory of the cave: What we see is but a shadow of what is really happening. That shadow changes shape, depending on the direction and source of the light casting it. So it is the height of folly to come to any conclusion about the *whole* after looking at just *one* of the shadows it projects.

Product	Monetary Value (millions of dollars)	U.S. Share of Japanese Imports
Aviation equipment and parts	1,447	97.5%
Computers	798	77.9
Internal combustion engines	710	84.6
Integrated circuits	426	74.2
Office equipment and parts	372	76.7
Electrical circuitry equipment	261	61.0
Photographic and film materials	256	74.8
Medical equipment	226	63.3
Parts for scientific optical equipment	212	79.8

Source: JETRO, *Quick Reference Handbook of Economic Relations Between the United States and Japan*

Figure 5. Major U.S. Products Exported to Japan (1985)

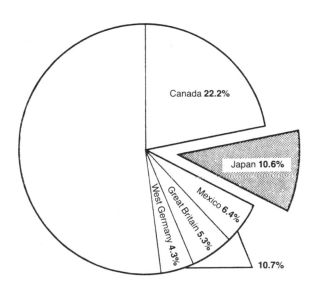

Source: JETRO, *Quick Reference Handbook of Economic Relations Between the United States and Japan*

Figure 6. Share of U.S. Exports Taken by Country (1985)

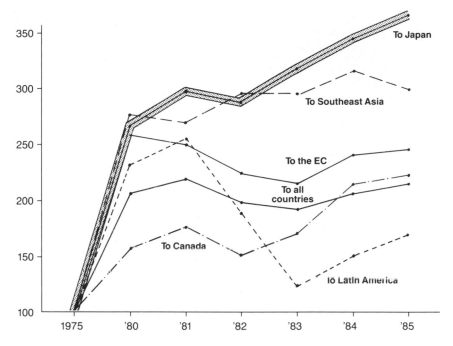

Source: JETRO, *Quick Reference Handbook of Economic Relations Between the United States and Japan*

Figure 7. Trends in U.S. Exports by Region (1975 = 100)

The old story about the blind men and the elephant is also apropos. Even people involved in U.S.-Japan trade negotiations have no better idea of what the elephant looks like than did the blind men. Even worse, most of the daily news we get tends to describe the dirt on the tip of the elephant's tail or the rocks at which the elephant is pawing. People use these to argue the whole. At times like this, it is crucial to step back and get a perspective on the problem.

A failure to gain this critical perspective underlies the attitudes and actions of both Americans and Japanese. A discussion I had the other day with an advisor to the U.S. Department of Commerce illustrates this well. Naturally enough, we talked about leading edge technology.

"We'd like to buy more American semiconductors," I told him, "but their quality is poor." And I gave him data to show that their failure rate was ten times higher than it should be.

He suddenly changed the subject.

"American communication satellites are of high quality. Why don't the Japanese buy them?" he asked.

"That's a different matter altogether," I countered. "We can't just go out and buy that equipment, because the allocation of frequencies for satellites in Japan is different from the allocation in the United States."

Before long, the discussion switched to beef, and when I said, "We're buying plenty of American beef," he responded, "No, no matter what you multiply by zero, the total will still be zero." I didn't really understand what he meant by that, and when I asked, he said: "The volume of beef exported by the United States has never been very high, and even if you buy 70 percent of it, that doesn't mean very much."

Now, that thought has occurred to me, too. On the international market the price of American beef is high. So should we go to the Ministry of Agriculture and suggest total liberalization? If that were to happen, I am convinced that not a single slice of U.S. beef would find its way into Japan. The U.S. Department of Commerce might find that acceptable, but I suspect that the U.S. Department of Agriculture would start complaining.

Ours was not, of course, a public discussion, and no doubt my host was speaking somewhat off the cuff. But even so, can we really expect this type of discussion, mired down in one detail after another, to further mutual understanding? A breakdown in communication is the more likely result.

Trade Friction and Contract Bidding

The problem here is one of perception. It is exacerbated further by the preconceptions of many Americans and a certain ineptitude in the Japanese response. These two aspects of U.S. -

Japanese trade friction were manifest in U.S. criticism of Japan on the question of the opening of the market in Japan for electric communications equipment. The sparks caused by this trade friction first burst into flame in 1980. At that time I worked for Nippon Telephone and Telegraph (NTT) and met Representative James R. Jones (D-Oklahoma), author of the "Jones Report," in Washington, D.C.

What he told me at our meeting was this: "What bothers us is the fact that NTT, the Japanese international telecommunications firm, uses optionally awarded contracts in the purchase of materials. We believe that it should institute the principles of free competition, with a bidding system."

I answered him: "You say that we have a closed system without competitive bidding, but our present system of optionally awarded contracts was instituted at the orders of General MacArthur when he was the head of the Allied Supreme Command in Japan. It's by no means a method invented here in Japan. And isn't the optionally awarded contract system the standard way of doing things in the West?"

He was surprised by this, and said, "That's the first I've heard about that." I went back to Japan and looked for any sources I could find from the occupation to document this point. I just couldn't turn up anything, not even at NTT, but in my search, I found out that the documentation I needed was in the possession of Sen Nishiyama, who had been an interpreter for the U.S. Embassy right after the war.

I immediately sent copies of these documents to Representative Jones and a number of others, and soon received a letter of thanks. The "Second Jones Report" came out shortly thereafter, and was considerably different from the first. It was so accurate that it almost appeared to be praising Japan. I remember thinking at the time, "That's a real strength of the United States. Top-level Congressmen are not like a lot of other people."

But there is another chapter to this story and that lies in Japanese response. As a first step in a campaign to neutralize American criticism that "NTT is a closed corporation," the company decided to purchase a quantity of pocket pagers made by an American company. In a case like this, it would have been logical to comply with American demands to use competitive bidding. In fact, the Japanese used the very optional contract system criticized by the Americans; they determined the companies they wanted to buy from and split the order among them. The Americans reaction to NTT's methods in this case was gratitude; it was recognized that American manufacturers probably would not have been able to land the contract if it had been subject to competitive bidding.

There are two lessons inherent in this story. The first concerns the criticism made by the United States. This began with a preconception, created a scenario, and then backed it up with a few concrete examples. In essence it relied wholly on theory.

The concrete examples behind such a theoretical argument need only be symbolic. An illusion is created from the very beginning; we eventually come to see that it is devoid of content. With this realized, the issue can be understood properly, and we can see that it never really was a major problem.

These principles also apply to the negotiations that took place at the undersecretary level in February and March 1985. On March 6, Japanese television carried an interview with a smiling Lionel Olmer, then U.S. Undersecretary of Commerce for International Trade. After Olmer returned to Washington, however, a hardline Representative, who may or may not have known the truth, started dumping on Japan in the usual manner. Although what he had to say was full of misunderstandings and prejudices, his remarks prevailed over the truth. As a result, Olmer's testimony was lost in the shuffle. His testimony on the real reasons for the U.S. trade deficit, such as the overvalued dollar and the inferiority of the American marketing techniques, was totally ignored.

After being beaten down like that, Olmer returned to Japan, where he totally reversed his position and took a hardline approach. At that point about all the Japanese negotiators could say was, "Good grief!"

But I have said that there are two lessons here. The second is the ineptitude of Japanese response. An official of the U.S. Department of Commerce, who had visited the Japanese Diet as an observer, said to me: "I have gone to Japanese bureaucrats with a lot of doubts I've had, and questions about procedures, and they always put me off — I never get a good answer. If these matters were under study, or if progress were being made, they should let me know what is going on, but they're always totally silent about things."

Under these circumstances, it is not really surprising that so many unnecessary problems crop up between the United States and Japan. In the strange times in which we live, all it takes is one thoughtless remark to ignite a crisis situation. These days emotions often hold sway over logic. Our overriding concern, then, should be to be sure that we have transmitted our ideas effectively. Otherwise we will continue to make stupid mistakes. In this last part of Chapter One, I offer my ideas on how Japan should shape its response to American criticism in a way that will benefit both countries.

Certification and Trade Friction

The problem of misperception that underlies much of U.S.-Japanese trade friction also comes to the fore in the matter of the Japanese certification system. The Americans say that the Japanese system of certification of products is overly picky — that any product which works without difficulty in the United States should be okay to use in Japan, as well. This dispute centers around communications equipment, but is by no means limited to that sector.

At first glance, it might appear as though Americans are correct in this assertion, but you cannot dispose of the Japanese certification system so easily. I have had considerable first-hand experience with certification. I spent two years as the head of a committee at the Agency of Industrial Science and Technology that studied international certification for electrical equipment. In 1984 I prepared a report on my findings. Our starting premise in the project was the idea that it was a waste of time for each country to have its own certification standards and that it would be better to have mutually recognized testing.

As our investigation progressed, we realized that it was not really all that simple. What we found was a whole host of factors. Each country has its own ways of doing things, users have their own habits and there are different ideas concerning safety. In a larger sense, the problem reflects cultural differences between countries.

Several examples will make this clear. Under the current system, water pipes for all public water lines in Japan must be made of bronze, and water authorities throughout Japan will approve no other material. In Europe, however, it is common to produce pipe from brass. Brass is cheaper than bronze. This became a problem several years ago, and the Agency of Industrial Science and Technology ran tests on the two types of pipe. These showed that brass would corrode if used with Japanese water, because Japanese water is soft, while European water is hard. Thus, the more expensive bronze has been designated for use in Japanese water pipes.

Switching equipment is a second example. While I was still at NTT, some of the switching equipment we had bought from the United States began to corrode while in storage, causing us considerable problems. The reason was that some of the material used in the insulation promoted the formation of mold. This type of problem would not occur in Japanese products because Japan has more problems with corrosion than nearly any country in the world.

These examples are but part of a fundamental misconception of the proper role of certification in manufacturing. After all, every country has the right to set up a certification system and to expect compliance with its safety standards.

The United States has its Underwriters Laboratories (UL) system: home electric appliances cannot be sold in the United States unless they carry the UL mark. The television sets owned by the members of Congress who complain about the Japanese certification system carry this mark. Washing machines, toasters, and vacuum cleaners will all have the UL mark as proof that safety standards have been observed.

All products must undergo expensive testing in order to be certified with the UL mark. Companies are subject to on-site inspection so that their production methods can be observed in the factory. The United States is not alone. In England, an even stricter standard known as the BEAB is in effect. There everything is checked, from the company's table of organization down to the standards observed within each department. The West German VDE system is even more stringent.

Why should Japan not be permitted to do likewise? When Japan attempted to establish a similar certification system for communication equipment, the Americans claimed that it would place their technical secrets in jeopardy. They maintained that only paper inspections should be conducted. This fact alone, in my opinion, is enough to show the ignorance of those who complain about certification systems. Furthermore, Americans claim that the Japanese certification system has been intentionally designed to keep foreign products out of the country. I would like for them to drop all these false charges for a while. In Japan we call this "making up a pretext," and any true gentleman would be too embarrassed to try it.

Japanese manufacturers comply with the certification requirements of other countries as a part of their worldwide export activities. We have a proverb in Japan equivalent to "when in Rome, do as the Romans do." It is only natural to follow the rules of the country you are in. The world does not exist for the United States alone.

So how do the Americans proceed with the certification problem? Since they hardly care about the content, they simply complain at random. If these complaints are not refuted on the spot, they make outraged comments to the mass media. This is nothing more than a game, where the object is to outscore your opponent.

Certification *is* related to quality. At the very least, the Japanese cannot feel assured about the quality of American products, until they have tested these products carefully. Shall I show you what I mean? I will content myself with a short example of something that actually happened. Anything not based on reality has little force.

As I noted above, at one point NTT decided to buy pocket beepers from an American firm as an expression of good will. Unfortunately, the beepers did not pass the first test they underwent. There were similar problems with the components in car telephones purchased at the same time. Several hundred of the mobile telephone units in the first shipment were unusable. NTT had been taking applications for subscriptions to this service, but was thus not able to fulfill them. Since we were in a rush, we decided to order this initial quantity from a Japanese manufacturer. But when they started to manufacture the units, it turned out that the phones' special specifications prevented the new components from being assembled simultaneously with those already available. By the fall of 1984, this incident had caused utter confusion among the manufacturers concerned.

The companies involved in this incident decided at the time to keep it as quiet as possible and did their best to keep it a secret. As everyone knows, the very act of hiding is to reveal; the fact that NTT had to purchase alternative products made it an open "industry secret." Anyone who knows anything about the backstage maneuvering that occurred in this case can only be exasperated at the gall the Americans display in their public statements on certification.

Beyond Certification to Parts Per Million

But exasperation by itself will not solve the problem. To cool my readers down a bit at this point, I would like to explain in more fundamental terms some of the differences in perception between American manufacturers and Japanese customers. Of course, these extend far beyond certification. One of the most important revolves around the concept of "acceptable quality level" or AQL.

Public data clearly show that American telephones break down between three and ten times more frequently than Japanese phones. When I mentioned this fact to an American congressman, he responded: "Then the Japanese just need more repair shops."

We can only take so much of such simplistic responses. It is no good for Americans to continue to ignore the quality control issue in selling their products in Japan. This is what is known in Japan as "hard sell." It will never be successful over the long term, and will leave only an increasingly unpleasant aftertaste.

The term AQL is widely used by American manufacturers in conjunction with the inspection of manufactured goods. It was first coined for materials purchased by the U.S. armed forces after the Second World War. Of course, it stands for "acceptable quality level" and has been translated as such into Japanese. For example, an AQL of 2 percent for a given product would mean that a shipment of that product could legitimately consist of up to 2 percent of defective goods. The AQL serves as a margin of error, since product inspection involves only a sampling. It is therefore not possible to guarantee a defect rate of zero and AQL serves as a goal for overall product quality standards.

AQL figures are also used in Japan. The way that they are understood, however, is completely different here. This has become another source of friction.

In the United States, if a given product run has an AQL of, say, 2 percent, then it is acceptable if up to 2 percent of the products in the lot are defective; Americans consider it a waste of time to try to keep that number below 2 percent. The Americans thus feel that it is rational to mix defective goods with the good products so that the shipment contains about 2 percent bad goods.

The Japanese, however, approach the situation differently. A customer pays for every product that he receives. That means that he should ask for a replacement for defective products. In other words, the Japanese believe that one should not be expected to pay for defective products.

The Japanese commercial code reflects these ideas. Statutes provide that the responsibility for defects lies with the manufacturer. The maker is responsible for any major flaws or defects discovered in products, even if the customer has accepted them after conducting his own inspection. These statutes were in effect even before AQL came to Japan from the United States. The idea behind them is that the manufacturer should assume all responsibility for his products. The idea of sending shipments that include defective parts is totally out of the question in Japan.

AQL has also been refined in Japan because of the intense competition among manufacturers. This guarantees that every effort will be made to totally eliminate defects, no matter what the AQL figure of a shipment. For this reason, more and more Japanese manufacturers have stopped measuring defect rates on the percentage (parts per hundred that are defective) level, and now use the so-called ppm (parts per million) standard. Obviously, products whose defect levels are measured in percentages will lose out to products where the measurement is done in ppm.

These differences in the understanding of AQL have led to continual trade friction. It seems clear enough that this difference is linked to the superior quality of Japanese products. For example, the quality of Japanese cars is guaranteed at the ppm level. This is an important aspect of their competitiveness.

There is no question that this problem was one of the biggest hurdles to be overcome when American manufacturers started doing business in Japan. The Americans immediately express doubts when they hear someone speaking in terms of "parts per million." They understand the concept, but feel that it is bound to be expensive to improve quality to that point; therefore, when the Japanese sell these products for less than American producers charge, it must be dumping.

This is the source of a serious misunderstanding. In fact, assuring quality at the ppm level rather than the percentage level is an effective way of cutting costs. This is amazingly difficult for Americans to understand because they can think only in terms of their own experience. They must first become aware that there are worlds within this universe with which they are not familiar.

The ppm method cuts costs because it drastically lowers the failure rate of products once they are in us. I stressed this fact as vigorously as I could during a seminar held in Washington,

D.C., in 1980 as a measure to help reduce U.S.-Japan semiconductor trade friction. At that event I tried to tell my audience that Japanese semiconductors cost less because they had a lower defect rate.

But they seemed to think that I was simply lying. Finally, though, I was backed up by David A. Garvin, a Harvard professor. Writing in *Harvard Business Review,* he offered comparative data for Japanese and American air conditioners: "The shocking news, for which nothing had prepared me, is that the failure rates of products from the highest-quality producers were between 500 and 1,000 times less than those of products from the lowest. The 'between 500 and 1,000' is not a typographical error, but an inescapable fact" (*Harvard Business Review* 61 (Sept./Oct. 1983), pp. 64–65). He states further: "The total costs of quality incurred by Japanese producers were less than one-half the failure costs incurred by the best U.S. companies." This, he notes, is because "failures are much more expensive to fix after a unit has been assembled than before," and Japanese companies had a considerable commitment to "design review, vendor selection and management, and in-process inspection" (p. 66).

Garvin's assertions provoked a considerable response in the United States and were discussed in major American business magazines shortly after the article appeared. Americans who wouldn't accept such facts coming from my mouth had to take notice when another American showed them the obvious.

How Should Japan Respond?

Earlier in this chapter I promised to offer some ways in which Japan should respond to the many accusations flung its way by the Americans. The causes of trade friction between the two countries are rooted in problems of perception, political posturing, and simple misunderstandings about the nature of economic reality. In order to eliminate these frictions, we here in Japan must tough it out on our own as well as work to create spokespeople for our cause within the United States. This is a good first step. Consider the following story in this connection.

The tale is about a unique man. His name is Masato Degawa. He was graduated from Oxford University in England, and became a fellow at the prestigious Matsushita Institute of Political Economics. He went to Washington, D.C., and became affiliated with the Brookings Institute. He then served as a member of the policy staff of the U.S. Senate Foreign Relations Committee. After completing his studies at the Matsushita Institute, he continued to live in Washington. Masato is an energetic young man and wants to improve the relationship between Japan and the United States.

In his capacity as a congressional staffer, he had the opportunity to view the actions of the U.S. Congress from the inside and actually become involved in its deliberations. He says that there is a group of U.S. Representatives and Senators who view Japan as the perfect pigeon in showing off their own power to their constituents. To gain political points without exposing themselves to harm, all they need to do is criticize Japan. The television broadcasters rely on them for information, and they appear in the newspapers all the time. This exposure ensures them free election publicity. In contrast, notes Degawa, the Japanese spend huge sums for lobbyists; this gets them virtually nowhere. Furthermore, the Japanese want to obtain all the information possible, but by the time they get it, it is already out of date.

Effective lobbying demands presence and connections. Degawa says that if he used his connection with Representative Stephen J. Solarz (D-New York) when he visited other members of Congress to get information, they would always answer his questions frankly and openly. Anytime he had to go without this type of connection, his source would hold back and tell him only a little of what he wanted to know.

Proper connections are not the end of it. You need to realize that each member of the Congress has his or her own area of specialization. The Japanese think that because a person is famous you can depend on him for anything. This is not true in Congress and there is no use in going to people for anything that falls outside of their own particular area. The Japanese who

does not understand this fact of life will get nowhere, no matter
how much money he has to spend on his lobbyist.

Stop Acting Like a Coward, Japan!

Basically, I am very fond of Americans. They are open to per-
suasion, frank, and openhearted. Once they grasp something,
they shake your hand without being too hung up on old ideas. I
found that both Representative Jones and Undersecretary
Lionel Olmer fit that mold nicely.

Unfortunately, when one looks at the details of trade friction, it is like peeling the layers of an onion. The American
with whom you are dealing will first think it strange when he
takes your hand. After he has looked into the matter, he will
often realize that the Japanese are neither plotting a conspiracy
nor planning to shut the door in his face.

Poor response to the media is another particularly noticeable
Japanese weakness. A prime example occurred in February
1987 when *Fortune* carried a special article describing spying activities allegedly being conducted by Japanese manufacturers in
California's Silicon Valley. The article was very flimsy and the
Japanese should have gone to court to demand a retraction.
They could have won the case. But the Japanese belittled the
idea of going to court, saying that people would realize the truth
nonetheless. The repercussions from that article continue to
linger. We are still suffering from the scars even now. If we had
gone straight to court when the article appeared and won the
case, we would have told the world that playing fast and loose
with Japan could result in someone's getting hurt. As the saying
goes, we could have "reaped blessings from disaster." It was a
major mistake.

Tomizo Arimoto, the former Japanese Ambassador to Norway,
wrote an article that appeared some time ago in the KDD (ITT)
house organ. In that article he related the following incident:

> One-third of the cars on Norwegian roads are Japanese made. Further-
> more, Japanese electrical appliances, cameras and the like have

flooded the country. Of course, Japan has a very large trade surplus with Norway. I was worried about that, and took it up with a high official in the Norwegian government. He told me not to worry and said: "No, it's only natural that Japan should be in the black in trade with us. But on the other hand, Japan is using a large number of Norwegian boats. Thanks to that, Japan has become the best customer for Norwegian maritime transportation. We need to thank you." He told me that the one thing that he couldn't understand was that Japan totally failed to respond when a certain superpower in the EC started a vicious Japan-bashing campaign.

The final comment on this story contains an extremely critical point that can be profitably considered in the context of the relationship between the United States and Japan. In 1985, the Japanese world trade surplus amounted to approximately $52 billion. This was exceeded, however, by $65 billion of long-term capital drain. Although the United States complained about the movement of goods, it was strangely silent concerning the movement of money for some time. In early March of that year, Paul Volcker, at that time chairman of the Federal Reserve Board, finally commented that investing this $65 billion within Japan would cause the country's GNP to grow and would take care of the trade surplus problem as well. He suggested that Japan stimulate a 10 percent growth in its GNP.

There may well be some truth in this, but surely it is not completely true! How did the Japanese government counter these remarks? It said nothing and let them stand unchallenged. The Japanese often express rather volatile opinions on domestic issues, but they make few effective pronouncements when it comes to international economics. I cannot help but wonder why there are absolutely no examples of a Japanese expressing an opinion that both floored the opposition and had a real impact on international relations. We could be called "wife beaters" — bold at home, but without courage when we leave the house.

Even when the Japanese message gets through, the Japanese often fail to drive it home. Follow-up continues to be a big problem. The quality control issue again comes to mind. The fact that Japanese quality control has been the force prying open world markets is now accepted everywhere. The study of Japanese methods is now commonplace. The driving force behind this awareness began with the 1980 semiconductor seminar held in Washington, D.C. Top-ranked American firms had just received an ugly shock with the news that American products had 10 times more defects than Japanese products! This was reported in May 1980 in an influential NBC White Paper called "If Japan Can, Why Can't We?"

Why doesn't Japan follow up in cases like this? A sustained effort is needed. Simply making chance remarks will have no effect. As one of the organizers of the Washington seminar, I felt that it was a real pity.

Japan could learn from other countries in this regard. Many go straight for the jugular to take advantage of America's weak points in business transactions. One famous example of this is Great Britain. During arms reduction talks between the United States and the U.S.S.R., Prime Minister Margaret Thatcher reversed her previous position and announced her total support of the American Strategic Defense Initiative (SDI). In return for this act of support she was able to have the American investigation of British Airways antitrust violations abandoned on the grounds that it did not violate U.S. law.

The Japanese would do well to heed a remark made by Ambassador Mike Mansfield in the November 11, 1983 issue of the *Japan Times*: "The Japanese need to curb their courtesy and reserve and speak out when they believe themselves to be right."

Kneejerk Responses Will Not End Trade Friction

In spite of the facts, Japan has borne the brunt of world criticism for its trade policy. The openness of Japanese markets is a case in point. It is clear that the Japanese markets are much more accessible than those of the EC. France, for example, has engaged in a campaign against imported videocassette recorders. In spite of this, though, Japan still receives the shrillest abuse. The explanation for this lies in the fact that Japan responds so poorly to criticism.

A 1980 seminar held in Washington, D.C.

I have a proposal to make in this regard that may improve matters all around. My proposal is that Japan align itself with those Americans who are pressing for a restructuring of the American economy by pointing out the precise problems and framing solutions. All informed Americans have recognized the fact that the high interest rates and the overvaluation of the dollar that have prevailed in the United States since the advent of "Reaganomics" have fed the American trade deficit. In a special edition of *Time* magazine issued in Japan in March 1985, one article pointed out that Japan had become the world's largest investor nation. The article contained no criticism of the closed Japanese market but noted that the overvaluation of the dollar and high American interest rates were responsible for the Japanese trade surplus. On the contrary, it noted that investment from Japan was responsible for putting the brakes on the rise in American interest rates. In the same vein, Japanese efforts to bring about appreciation of the yen were highly praised.

In order to reach the root of the problem, Japan must do what it can to help the United States restructure its financial policies. All possible analyses must be attempted. Some will no doubt claim that Americans will not permit this as a matter of pride. But most Americans are sincere. We need to join forces, establish a consensus, and then put the policies into effect.

It is obvious that economic relations between Japan and the United States (or between any two countries) are composed of a large number of fine points that cannot be understood in terms of a simplistic "fair vs. unfair" view of the world. The American habit of holding just one small aspect of Japan before a magnifying glass and then coming at us with a sword raised to deliver a death blow fails to persuade; all it does is leave one with a sense of the poverty of the thinking behind it.

President Reagan said something that I think was very good at a November 1983 palace banquet while he was visiting Japan:

> Combined, the gross national products of the United States and Japan come to over one-third of the world's GNP. Further, the leading technology of the world is concentrated in these two countries. So by joining hands, these two countries should be able to light up the hard-to-find exit from the world of darkness and confusion.

I am in complete agreement with these words.

It is not that I totally fail to understand the reasons that the United States has shot one critical arrow after another at Japan. I think they arise from a basic cultural difference that exists between the two countries.

Japanese people who have lived abroad for any length of time will understand what I am talking about. If a Japanese should happen to have a slight traffic accident or a simple misunderstanding when shopping, he will tend to shrug it off with a simple "excuse me." Americans, on the other hand, rarely admit their culpability. They tend to rationalize the situation and transfer the responsibility elsewhere. Some people claim that this difference derives from farming and hunting cultures; there is no question that these same tendencies can be found in business and political dealings as well.

The following story was told by a friend of mine who recently returned from Singapore:

> I went out for a drive one day, and collided with a car that had come across the center line. The accident was clearly the fault of the other driver. We talked it over, but couldn't reach any agreement. This bothered me, and I wanted to call the police and let them take care of it, but the other driver said, "They won't come when you call — it would be a waste of time."
>
> I got sort of mad at this, and finally found a beat cop. But as soon as he found out that no one was hurt, he just said, "I'm busy — talk it over between yourselves," and did nothing. I decided to give up, and started home, and when I did that the other driver suddenly got bossy — "Hey, I've just lost 30 minutes! How are you going to repay me for them?" Then things really got bad.

Until they are fully acquainted with local conditions, Japanese always come out losers in situations like this. Every time I look at the current trade friction problems, I inevitably think of this story.

At this point I'd like to draw your attention to Figure 8. It depicts the trade revenues for a number of countries. As you can see, the present severe imbalances began to develop after President Nixon's visit to China and the two oil crises that followed that event. These latter events were severe economic shocks to most countries in the world, but it was Japan that rebounded from them most successfully. Recovery did not proceed smoothly for the United States or for European nations.

Without fear of exaggerating, America's propensity to blame others for this slow recovery can be called the real reason for its criticisms of Japan. But blaming others for the deterioration of its economy will not help the United States solve its problems. U.S. Ambassador to Japan Mike Mansfield, again in the November 11, 1983, issue of the *Japan Times* made some excellent remarks in this regard: "It will be of absolutely no help for the United States to criticize others for its own lack of economic vitality. The economy will not begin to improve until the U.S. searches within its own shores for the causes of the problems, and makes a continuing effort to eliminate these causes."

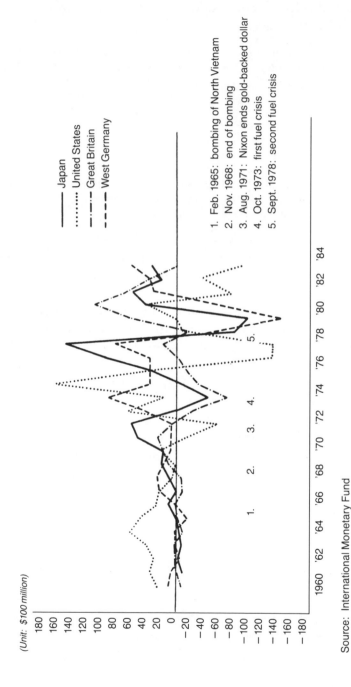

(Unit: $100 million)

Source: International Monetary Fund

Figure 8. Normal Revenues of the Advanced Industrialized Nations

In other words, as long as it clings to its geocentricism, the United States cannot expect any relief from its economic problems. The oft-used expression "equal partners" should mean that the Japanese may be allowed to continue to travel on the left side of the road with no obligation to change to the right side. Those who are aware of the true nature of U.S. -Japanese trade friction understand this well. These are words that should give us courage. As Japanese, we must make an effort to respond to them in a realistic fashion.

First, though, we need to cast away political trappings and come to an understanding of the situation based on an objective understanding of reality. No reasonable conclusion will ever be reached if the United States and Japan continue to obsessively drag up one difference after another and trade insults. The solution will come only through hardnosed analysis and hard work.

CHAPTER 2

The Hollowing of American Industry

In the United States you sometimes hear the expression, "Don't shoot yourself in the foot." In manufacturing it is used to express the sentiment that America has gone to a great deal of trouble to develop its high tech industries, and they should not be moved overseas just to reduce costs. The feeling is that moving such industries will serve only to strengthen the industrial power of their new sites and eventually those nations will set their sights as independent competitors for U.S. markets.

People in the United States criticize the Japanese market for being closed and say Japan is flooding the United States with subsidized exports. Even if all barriers were removed and all exports of Japanese cars to the U.S. were halted, the effect on the gross national product of the United States would be less than 1 percent. If American industry rushed to establish overseas bases the effects would be much more devastating. It would mean, in effect, the total collapse of American industry. Staving off Japanese exports and opening Japanese markets are minor matters compared to this distinct possibility. The basic problem today is not to move production overseas but to strengthen American industry at home and make it capable of standing on its own legs.

Stabilization of American manufacturing must occur behind the artificial protection of import allocations. We need only to look at the steel industry to see how true this is. Worse, industries one after another seem bound to repeat the failures of the steel industry. The American television industry has gone by the boards, for example. Even the fates of the semiconductor and personal computer industries, showcase segments of the high tech industries, may be in doubt over the long term.

On July 5, 1985, American Telephone and Telegraph (AT&T) announced the layoff of 875 workers. These jobs were lost because AT&T shifted its manufacturing base to Singapore. This news is symptomatic of American deindustrialization. The telephone, invented by the American Alexander Graham Bell, has been one of the cornerstones of today's information society. AT&T is a giant corporation. Thus, the news was a considerable shock to the American public. Once the production of telephones has been shifted abroad, it seems safe to assume that it will never again return to the United States.

Similarly, virtually all components for personal computers made in the United States are imported from abroad. Even industry giant IBM relies on foreign production for some 70 percent of the structural components of its personal computers, according to a recent article published in *Business Week*. America believes itself to be world leader in leading edge technology. Nevertheless, the personal computer, itself the main symbol of that supremacy, is in major trouble.

The overvaluation of the U.S. dollar is, of course, one reason for the rush to overseas. It is only natural that American industry should rely on overseas production with the dollar at a substantial premium over its true value. Thus, more than the shortsightedness of American managers is to blame for the situation. American industry's movement abroad will bring down the exchange rate for the dollar and continue the downward movement in American capabilities.

The United States trade deficit is not with Japan alone. For example, in 1984 the Taiwanese trade surplus with the United States totalled $11.1 billion, while the Japanese surplus was $36.8 billion. When we consider that Taiwan has only 18 million people, the Japanese surplus becomes relatively insignificant.

Despite the daily complaints that Japan is not buying American goods, Japan still rates second as the purchaser of American exports, after Canada. This can be seen from Figure 9, which shows the import and export relationships between leading nations. In 1984, Japan bought $21.8 billion worth of U.S. products. As stated before, Japan is the largest purchaser of U.S. agricultural products, having bought 17 percent of U.S. agricultural exports in 1984! In fact, America should be thanking us.

But I do not really want to reiterate my emotional opinions here. There are more important matters at stake. If things continue on their present path, the United States, which has had the strongest industrial capacity in the world, will be in danger of industrial collapse. I think so-called world leaders should be giving some thought as to how to prevent this from happening. This issue is far more important than debates about the opening of the Japanese market; it has a significant bearing on the fate of the entire free world.

There will doubtless be some readers who will say that stating the matter in these terms will not only fail to penetrate the consciousness of prideful Americans, but will actually be counterproductive. Well, Americans are human like the rest of us, and some of them would probably react in that fashion. But everything that I have said is the truth. I have distorted nothing. Reality must be perceived directly, just as it is, even if it is not very pleasant. Fortunately, there is now a smattering of people in the United States who realize the dangers associated with the exodus of industry and who believe that something must be done about it.

The Ceaseless Drain of Industrial Power

The hollowing of American industry proceeds at an ever greater pace. Statistics from the Bank of Japan show that between 20,000 and 30,000 American industrial workers lose their jobs each month. These are terrible figures. In Japanese terms, this is equivalent to one major corporation going bankrupt *each month*. U.S. service sector employment is growing,

Importing country → / Exporting country ↓	EC	U.S.	West Germany	Japan	U.K.	France	Industrial countries	Oil producing countries	Non-oil producing countries	Communist countries	World totals
EC	299,273	44,232	70,671	6,487	41,856	59,107	422,602	46,155	81,068	15,983	574,496
U.S.	44,311	—	8,737	21,894	10,621	5,961	117,617	16,419	58,949	2,595	200,535
West Germany	81,522	12,797	—	2,189	13,879	21,834	125,810	12,451	23,489	6,608	169,436
Japan	18,529	43,339	5,882	—	4,984	2,010	74,115	19,411	43,934	3,868	147,000
U.K.	40,118	12,679	9,164	1,208	—	8,527	67,034	8,584	14,311	1,247	91,639
France	44,907	5,713	14,197	1,087	6,913	—	61,642	8,396	17,612	3,204	94,945
Industrial countries	430,529	152,026	105,518	40,638	75,359	76,198	779,922	93,192	210,001	32,195	1,139,900
Oil producing countries	48,844	21,546	9,697	40,350	3,601	11,940	122,281	3,482	57,485	656	192,700
Non-oil producing countries	66,319	69,698	16,788	27,581	12,562	11,096	183,014	22,531	79,190	21,953	323,626
Communist countries	20,577	683	6,110	1,675	1,655	3,352	31,048	1,956	17,724	—	50,728
World totals	568,685	245,192	138,936	112,745	93,781	102,802	1,124,000	122,936	369,363	54,803	1,721,928

Note: Based on customs statistics (FOB), and does not include trade within the Communist Bloc nations
Source: Bank of Japan, *Comparative International Statistics*

Figure 9. World Trade Matrix

but some people have even begun to argue that at the current rate the number of people employed in secondary (manufacturing) industries will drop to 20 percent of total employment.

In any event, more than anything else, the secondary industries keep American industry going. In the service sector, productivity is low; wages are only about 70 percent of those in manufacturing. It is true that external factors have caused these cavities to develop. Exchange rates and the successes of the newly industrialized Asian nations, including Japan, are among these. But I feel compelled to point out here that internal problems existed long before these; they center on the general posture of management in the United States. This is proven by the fact that Japanese manufacturers can make money with production in the United States.

Color televisions are a good example. More recently, Korean manufacturers have started to move manufacturing bases to the United States. American manufacturers of color televisions, on the other hand, have virtually all gone abroad. They either rely on OEM (original equipment manufacture) supplied from "offshore" countries such as Mexico, Taiwan, and Singapore, or they import products manufactured at plants they directly manage. This is a very strange situation indeed.

I also mentioned earlier that the same trend is starting in the automotive industry. One Japanese automaker after another has opened manufacturing facilities in the United States, a fact that has become the talk of the world. But Ford is now building a large plant in Mexico. There is no guarantee that the same thing that happened to color televisions won't happen with cars.

If this isn't unfair, I'd like to know what is. *Japanese industry* is working as hard as it can to protect American jobs. At the same time American industry, lured by the profit motive alone, is taking steps that can only result in a loss of jobs. This is something like pouring water into a sieve. I've finally figured it out — it's "unfair" if someone else does something bad for the United States but "fair" if the United States does the same thing.

If Japanese industry can make money manufacturing in the United States and American industry cannot, there has to be something wrong with American management practices.

One of the most recent topics of conversation is the news that the American high tech industries have started moving to Mexico in search of lower labor costs. Mexico has a large number of severe restrictions on foreign capital, and inevitably requires the investment of private capital. Even so, however, we have had news that IBM is constructing a wholly owned manufacturing plant there.

The losers in this situation are American workers. As soon as new technology is developed, employment possibilities are immediately shifted abroad. Atari, one of the leaders in the personal computer industry, gave up on American production and moved to Singapore as early as 1983. This resulted in layoffs of 1,200 people. This even preceded the abnormal strengthening of the dollar.

In order to survive today's competition, American manufacturers have given up all pretense of appearances and are willing to go anywhere in the world as long as they can manufacture their products in the most profitable location. Their special managerial forte consists of cleverly increasing profits through these movements. It is no wonder that the unitary tax scheme has been created as a way to assess the overseas profits of these firms.

I am not opposed to offshore production per se. Japanese manufacturers are doing it too. They have gone to parts of Southeast Asia in search of lower labor costs. This is really a matter of extending prosperity to a variety of local societies on a mutual basis. If anything, it will be more important in the future than it is now. The logic that dismisses these considerations and attacks Japan solely on the basis of its trade figures never fails to get me mad.

This illogic extends to the calculation of trade figures themselves. The electronics industry is a good example. A number of U.S.-based electronics manufacturers have established factories in Japan and produce semiconductors in a big way. IBM Japan is the largest exporter of computers from Japan. Many other firms engage in similar trade. We shouldn't overlook the fact that these products are included among the total figures when people speak of Japanese exports. The *Economist* offered an interesting insight in its issue of July 5, 1985. Calculating imports on a per capita basis — defining all products made in Japan by firms capitalized in the U.S. as "American," and products made in the U.S. by Japanese firms as "Japanese" — leads to a startling conclusion.

The Japanese spend $480 per capita on products imported from U.S.-based companies, while the average American spends only $287 on products made by Japanese companies. Looked at in this way, America has an overwhelming export surplus in its dealings with Japan. This is only natural since the population of the U.S. is twice that of Japan. If citizens of each country were to buy manufactured goods from the other in accordance with their buying numbers, the overall national purchasing power should mean that Americans would spend about twice as much on Japanese goods as the Japanese do on U.S. goods. On this proportional basis alone Japan would have an export surplus even after correction of the overvalued dollar. Given the above, it is strange indeed that the U.S. should bear a grudge against Japan at all.

Holding the Power of Life or Death over the U.S. Semiconductor Industry

Let's get back to our main subject. Here is the scenario in brief. U.S. industry has started setting up factories all over the world, without regard to location. This has weakened the competitive power of certain high tech industries in the U.S.; the semiconductor industry has been the main victim. American managers have been "shooting themselves in the feet" with their cavalier attitude. A little history is in order.

The key product of the semiconductor industry is the LSI (large-scale integrated circuit). These consist of several tens of thousands of transistors mounted on a silicon substrate only a few millimeters thick. The technology used in the fabrication of these devices was originally developed as a part of the American space program. After the end of the Apollo program, however, orders from NASA dropped off and Silicon Valley was plunged into a depression. Highly respected Ph.D.s could be found driving taxis. Their salvation came from Japan in the form of a high demand for pocket calculators.

At that time, calculators were composed of integrated circuits (ICs). The size of ICs meant that only a few hundred transistors could be mounted on a single chip. The fabrication of a calculator demanded nearly 100 chips. This meant in turn that the final product had to be the size of a briefcase, with a price to match. At this point Japanese manufacturers turned their gaze to LSIs being made in the United States. There was great excitement — these devices would allow a calculator to be made with only one chip! They were a bit expensive, but their size outweighed their price. Japanese manufacturers immediately opened offices in Silicon Valley, and began placing big orders for LSIs. While LSIs for NASA had numbered in the tens of thousands, this new use resulted in orders well into six- and even seven-digit figures. The new business revitalized the American semiconductor industry. In plain terms, the firms in Silicon Valley owed a new lease on life to the Japanese calculator manufacturers.

This corresponds to the second energy crisis in 1978. As part of their strategy for dealing with this, many American manufacturers decided to move labor-intensive assembly processes "offshore" to Southeast Asia where labor was less expensive. Chips were loaded onto planes and flown to overseas plants for assembly, then shipped home by air as completed products. At first glance, this way of proceeding appears rational, but it created its own set of problems.

Japanese manufacturers, on the other hand, introduced *automation* to the assembly processes in order to resist rapidly escalating labor costs. Automation is a more effective response to high labor costs. A human being can't really hold a candle to a robot. No matter how much of an advantage you might get from low wages, there are still tradeoffs: when it comes to precision work, no one can match the production uniformity of robots. Furthermore, robots can work 24 hours a day without tiring.

The way Japanese manufacturing has used automation accounts in part for the differences in quality and productivity between Japanese and American products.

Robots have another ability that makes them crucial in semiconductor production; unlike humans, they work very cleanly. The production of semiconductors requires accuracies down to one-thousandth of a millimeter; this makes it a struggle against contamination. The demands are strict. Air contamination must be less than 100 grains per 30 cubic centimeters of air. There has never been any question that jobs with these low tolerances can be done infinitely more precisely if fully automated.

With automation and sound management, Japanese firms began to dominate the semiconductor industry. This domination started with the production of RAM (random access memory) devices. The Japanese share of the 16 kilobit RAM devices was at one point 40 percent.

In the 64 K class, the Japanese share rose to 70 percent in 1982. In the 256 K RAM market, the predominate product group, the Japanese share has now become 90 percent. At each of these stages American manufacturers were, in the final analysis, overwhelmed by the strength of Japanese products. The reasons for their defeat are simple. With the exception of a few outstanding firms, production flow of American firms is poor. In addition, the failure rate for Japanese products is far lower than that of the American products. The failure probability for Japanese RAM devices is .0000001 per hour of operation — the equivalent of one failure per thousand years of operation. This performance is high even by American military standards.

The Japanese market share in semiconductors was achieved without dumping and without formation of cartels. It goes without saying that there was no collusion between the government and the private sector. If Americans are unable to meet the demand, they will not increase their market share. It's as simple as that.

The new stage of development in semiconductors is the 1 megabit device. Japanese manufacturers had begun shipping samples by late 1984, and most people believed they would capture this market, too.

But in July 1985, the Semiconductor Industry Association, an American trade association, filed an unfair trade complaint against Japan with the U.S. Trade Commission, and an investigation was started. What, one wonders, was the real goal of this action? At the time, American computer manufacturers had already launched new designs based on the samples they received from Japan. If the supply from Japan were to dry up, these manufacturers would be unable to make new computers. It would be extremely difficult for the American companies to design competitive new computers if the Japanese computer manufacturers were the only ones allowed to use the latest 1 megabit RAM devices in their new product designs. Americans would be stuck with outdated earlier products. Talk about shooting yourself in the foot!

The semiconductor is a basic staple of modern consumer and industry products. Banning imports from Japan would have a tremendous effect throughout American society. Everything from computers to electronic equipment, from aviation devices to machine tools, would disappear.

Reality shows us that the solution does not lie in slowing down exports from Japan or in forcing Japanese companies to build factories in the United States. The problem that requires action is much more fundamental: it originates in American management practices.

The Japanese have been successful in the semiconductor realm, but this is nothing to brag about in particular. On the contrary — if American agricultural exports to Japan were to suddenly stop, Japan would be thrown into utter turmoil. I am sure that many of my Japanese readers remember the horrible confusion that resulted in the past when U.S. soybean exports were just slightly reduced. Similarly, semiconductors have become the world's main industrial staple, and Japan has become the most powerful supplier of these devices.

Trade between nations has always depended on mutual relationships — ties that cannot be severed, no matter how much we hack away at them. For example, Japanese camera lenses lead the world market today, but I wonder how many are aware that France is virtually the sole supplier of the rare earth elements so essential to their manufacture. France developed the necessary refining technology while still a colonial power in Africa and has continued in its role as principal supplier. If France got out of sorts and stopped exporting this material to Japan, it would mean the end of Japanese production of high-performance lenses.

The grown-up way of dealing with trade problems is to restrain our emotions when we get a little angry and find a way to ensure mutual prosperity and survival. The resolution unanimously passed by the U.S. Senate on March 28, 1985, was a pretty poor way of dealing with the trade tensions that existed at the time. * It approved Japan-bashing with no opposition.

It has always been strange for any type of resolution to pass without opposition in the U.S. Congress. Not even the resolution that propelled America into World War II was passed without opposition. Opposition has been the conscience of America. The fact that this resolution was passed without opposition certainly shows that the situation is unusual. What would have happened, one wonders, if someone had proposed this type of resolution against any of the European countries? It would probably not have been such a simple matter.

The European countries all have markets that are more closed than the Japanese market. France's war against video equipment imports, for example, will go down in history as a very strange event. Earlier, when the American firm GE attempted to acquire Bell, the French computer manufacturer, the French reaction was also rather amazing. Nevertheless, we didn't hear the Americans screaming "unfair" at the French as loudly as they say it to the Japanese. Some people see this as being caused by a resurgence of the "Yellow Peril syndrome," but this is not right. The yelling at Japan comes because Japan won't yell back logically and loudly.

---

* S. Con. Res. 15, passed 92 to 0, was a nonbinding concurrent resolution that asked the president to restrict imports to Japan by an amount equal to the increase in automobile imports from Japan following the end of the voluntary restraint agreement on March 31, 1985, unless Japan took steps to open several key markets to U.S. trade.

According to the Japanese perspective, it is possible to insist on one's own sense of self-justification without using rhetoric designed to belittle the other party. Any number of examples of this can be seen by simply observing the United Nations in action.

The Greek statuary on display in the Louvre in Paris was made by people of a culture that no longer survives. Those who do not press forward firmly with their own claims to existence will soon be eliminated from this world. People from countries where this way of thinking is prevalent doubtless view Japan's apparent lack of self-justification as being somewhat strange. This finds its most extreme manifestation in the current trade friction problems.

Fortunately for us, some of the leading newspapers in Europe and the United States are beginning to carry opinions that run counter to illogical American assertions too frequently heard. Coverage of a recent controversy concerning Japanese retail practices is an example. Criticisms of the Japanese distribution system, coupled with demands to loosen the regulations placed on large retail stores, were made by some elements in the United States. These were tantamount to demanding that Japanese culture itself be changed. Such demands should not be met. One American economic newspaper was kind enough to write that an American-induced suppression of the one million plus stores in Japan would have a backlash that would reverberate for several centuries into the future.

American Managers Must Get Back to Basics

There is only one solution to ensure the future of American industry. American management must get back to the basics. This, and this alone, will allow the building of a prosperous economy. Despite this obvious need, American industrialists have done just about everything else in their efforts to keep afloat. They have seesawed between the "money game" and crying "unfair" whenever superior Japanese management has caught them unaware. The development of VCRs is a case in point.

The first company to actually make a VCR was an American firm. Ampex Corporation built the original machines for use by broadcasting stations. They revolutionized that industry. Their cost, however, exceeded $100,000 and there was no way they could be sold to the average consumer. Manufacturers throughout the world thus began searching for ways to produce them at a lower price for use in the home. This required a cost reduction that would allow them to be sold at about 1 percent of their original price. A number of would-be manufacturers failed in this endeavor, but finally a Japanese company succeeded in producing a viable product — the Sony Betamax. The value of Sony's annual production of VCRs went on to exceed $100 million.

What were the Western manufacturers, with their great technical and capital resources doing during this time? RCA, with its huge patent revenues from the development of television, acquired Hertz, the car rental agency, and became absorbed in further diversification. By the time it returned to its senses, RCA was in the position of having to rely on Japanese companies for virtually its entire supply of products. In Europe, Philips and a number of other companies attempted to resist the Japanese advance by developing a different type of video recorder. This culminated in a new low in business stupidity — a modern-day battle of Poitiers. Although the European companies attempted to stop Japan's advance into their markets, they were ultimately powerless to do anything other than bow to the conqueror.

Shouting "unfair" at every Japanese move in the development and marketing of new products and indulging in rampant protectionism is not only illogical, it may be counterproductive. The fact is, when it comes to market share, if one side wins, the other loses. If it were a matter of frankly admitting this and of giving the weaker party some type of equalizing handicap, then no one would be able to get angry about things. The practice of talking as though Japan were the only villain in the crowd resembles nothing less than simple foul temperedness. To scream "unfair" at others while disbanding your own industry is truly illogical. It's hard to take.

So I want to communicate this message loudly to American management: Return to the spirit of the industrialists. In this world the word "unfair" is found only in the vocabulary of losers. Americans *can* be winners in Japan and there are many examples. A big topic in Japan these days is the fact that Schick has captured a 70 percent share of the Japanese razor market. The top-selling McDonald's hamburger franchise in the world is in Kamakura, Japan. Never forget that, with a little persistence, the Japanese market will inevitably pay off.

European products can also do well. BMWs sell very well in Japan. Similarly, a particular West German precision machine tool is used widely in the factories of Japanese VCR manufacturers. This tool enabled the Japanese to manufacture the VCRs that have taken the rest of the world by storm. Any good product will always sell in Japan. Shoddy products will never sell.

There is no major difference in the type of equipment used in Japanese and American factories, whether for the production of cars or semiconductors. Nor do Japanese workers perform any better than American workers. The products made in the two countries are identical in principle. So what, one might ask, is the source of the differences in productivity, in quality, and in cost?

Think about it this way. When Japanese firms move to the American market they are able to adjust their costs accordingly, but the only way that many American firms see fit to survive is to move to overseas production. This leads to one inescapable conclusion — the problem is with American management practices.

Though this message may not penetrate the consciousness of the sorts of antisocial auto executives who take advantage of import restrictions to pocket large bonuses, I have great hopes that more socially aware Americans will begin to advocate a new policy of reindustrialization.

CHAPTER 3

The Secret of Japan's Production Strength

The bill's come due for the big money game. Anyone would feel bad after being beaten out by another manufacturer in the market. Even so, we Japanese can't understand the psychology of Americans — why do they waste their energy trying to influence politicians instead of concentrating on developing products whose excellence will allow American manufacturers to compete head-on with Japan and other competitors? This is derived from a difference in management philosophies. These differences, in turn, are rooted in the general customs in the United States; among the customs is a fascination with what I call the "money game."

The other day I talked to the president of a company in Nagoya that makes gas appliances and equipment. The gist of the conversation went as follows. The Nagoya firm formed an unexpectedly successful joint venture with a company in New Zealand. But an unpleasant incident occurred at a recent meeting of the board of directors. One of the New Zealand shareholders in the company had sold his shares to Merrill Lynch in New York, and Merrill Lynch now held 20 percent of the stock in the company.

At the time, the company was retaining a considerable amount of its resources as reserves in anticipation of expansion of the plant. The Merrill Lynch representative at the board meeting opposed this policy, arguing that the funds should be distributed as dividends to the shareholders.

A scene from the New York Stock Exchange

The president of the company balked at this demand, but the representative persisted. "I recommended that we buy stock in this company because of your high level of reserves," he said. "If you don't pay this out as dividends, I'll get fired — so do something about it!" The argument went on, but the president held firm and would not release the reserves as dividends. "This is why America is going downhill," he said to me later.

In the American economy the money game takes precedence over all else. This concept underlies the SEC disclosure system that mandates quarterly disclosure of a firm's financial data to the stockholders. This single-minded obsession with short-term profits is gradually causing the hollowing of American industry.

The machine tool industry is a perfect illustration of this problem. Until very recently, the United States dominated the industry. Figure 10 shows that in 1983, the production of machine tools in the United States was valued at $1.6 billion. It ranked fourth among the world's machine tool builders. At the time, Japanese production was valued at $3 billion and the industry ranked first in the world. The decline of the machine tool industry in America is directly linked to the money game in that country.

(Unit: $1 million)

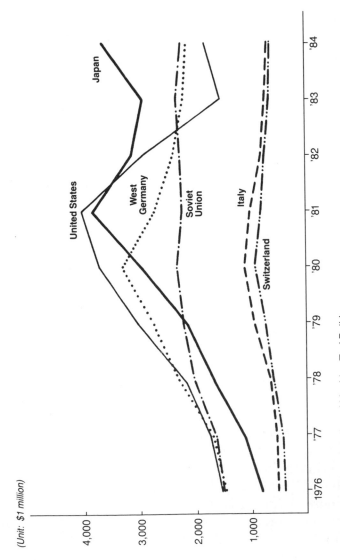

Source: Japan Association of Machine Tool Builders

Figure 10. World Machine Tool Production

Around 1979 there was a boom in machine tool production in the United States. Machine tool companies began to turn large profits, which made them very attractive as investment properties. American conglomerates began a series of takeover bids as a result. Like their counterparts in Japan, American machine tool builders are generally not large firms. Thus it is easy for someone with enough funds to hijack them. Unfortunately, the boom did not last long, and by 1982 the same conglomerates that had bought up the small machine tool companies began to divest themselves one by one of their unprofitable ventures, taking with them the capital assets these businesses needed to grow. A number of firms closed down their manufacturing divisions in the United States and kept their trademarks viable by importing goods from abroad. In short, they became brokers rather than manufacturers. This is a classic example of how the ever-popular money game has destroyed an industry in the United States.

Figure 10 charts the sudden and violent decline of the American machine tool industry. America is not alone, however; the problem has afflicted other countries, including West Germany and the Soviet Union. The machine tool industry is critical to manufacturing in general and its decay had ramifications extending into all areas of industry.

The situation has advanced to the point where it now affects national security. There has been talk of trade sanctions against Japan, of course, but also against West Germany, Switzerland, and other nations that export machine tools to the U.S. But the American machine tool industry has not been done in by lower prices from competitors — American manufacturers simply lowered production on their own accord.

Given this situation, it is hard to believe that anyone could think that import restrictions could possibly have any effect. Furthermore, it would be absurd to impose such restrictions where domestic production no longer exists. There are several models of American machine tools no longer in production; if imports were restricted the country would be like a weak victim with a knife held at its throat. This security threat has forced the United States to give up the idea of restrictions.

The hollowing of American industry is not limited to the machine tool sector. The manufacture of steel has an even longer history of decline; it provides a textbook illustration of the weaknesses in American management skills and emphases. Rather than follow the world trend toward coastal steel mills America stabilized its steel industry through the use of import duties and allocations. In the process the actual *production* of steel has dropped precipitously and U.S. firms are mere processors of this vital material.

The *manufacture* of steel is one of the fundamentals of the steel industry. This basic process has changed radically in recent years. While mills in Japan were switching rapidly to the use of oxygen, U.S. steelmakers kept on making steel the way they always had, with open-hearth furnaces. A convertor can be fired up in 15 or 20 minutes; it takes an open-hearth furnace a leisurely five to seven hours before it is ready to work. The difference is starker than that between steam-powered and sail-powered ships. As a result of this, the American steel industry has fallen into a state of decline as quickly as if it had rolled down a mountain path. It has *hollowed* to the point where virtually no blast furnaces — necessary for making steel from ore — are in operation. U.S. mills are mere processing houses that perform such operations as rolling and extracting steel acquired from overseas. Production of the high-tensile steel crucial to the automobile industry has dropped way off because converters are necessary to its manufacture.

At a conference the other day I gave a piece of my mind to a high-ranking American official who was complaining about access to the Japanese market: "Machine tools and steel are the basis for all manufacturing," I told him. "Just what do you intend to do if a crisis situation develops in these fields? If America doesn't do something soon, the security of the whole free world may be in danger."

An awareness of this danger has become evident among some attuned Americans. In January 1985 the President's Commission on Industrial Competitiveness, headed by John A. Young, president and CEO of Hewlett-Packard, released a report concerning policies designed to restore the competitive power of U.S. industries. Entitled "Global Competition: The New Reality," the report includes a graph that clearly shows the basic reasons for the rampant "hollowing" of U.S. industry. The major cause cited was the pressure being brought on industry from the money game.

The Inherent Strength of Manufacturing

Look at Figure 11. It shows how profits from manufacturing activities in the United States have fallen while profits from the money game have increased; as you can see, as recently as 1980 the two sectors were reversed. At one time, money was made in the United States by making things through the sweat of one's brow; today a single telephone call can unleash a torrent of profits. A variety of reasons have caused this state of affairs, among them American fiscal policies that have resulted in high interest rates and the relatively low status assigned to technical development. One thing is clear: If this trend is allowed to continue, the competitive power of the U.S. manufacturing industries will drop to an even lower point.

Money has value when backed by physical productivity. Money not backed by this type of production is no better than wastepaper.

Who are the people responsible for the emphasis on the money game? Some people hold that the outstanding students from America's business schools have been responsible for weakening that country's manufacturing activities; I believe that this is probably true. Such people spend all of their energy considering revenue statements, interest returns, and the like, and believe that effective management practices consist of the skillful manipulation of these factors. When these people come to positions of authority in manufacturing, they are bound to implement policies that will bring about the kind of "hollowing" so visible in the machine tool and steel industries.

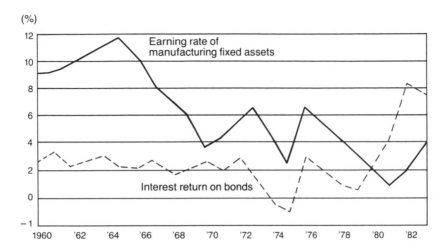

Based on study by U.S. Department of Commerce, Federal Trade Commission

Figure 11. ROI for Manufacturing Industries (Profitability of Net Assets) and Interest Return on Shares (1960-1983)

This is called the "softening" of industry, and is often spoken of in glowing terms, but the Young Report warns that it could have disastrous effects. Manufacturing is the key to maintaining a high standard of living. Productivity in the so-called third sector (service industries) is low. Wages average about 70 percent of those in manufacturing. An erosion of the manufacturing sector will therefore cause a decline in overall U.S. productivity and a corresponding drop in income levels and the standard of living.

Business Week published a special section on this topic, entitled "The Hollow Corporation" (March 3, 1986). This issue carried one headline that went straight to the point and with which I wish all Japanese intellectuals were familiar: "The Service Economy: No Paradise." There are those who tend to classify all industries that aren't directly engaged in the manufacture of products as a single, amorphous "soft industry." This is an oversimplification. The jobs of a waitress who hands out food in a restaurant and a computer programmer are completely different. Treating the two as though they were equivalent will give rise to a whole host of misconceptions.

Accountants, doctors, teachers, and the like are all included as a part of the service sector; computer programmers are not. As anyone who has attempted to write his or her own computer program will know, the task is little different from designing cars or gathering and analyzing data concerning optimal operation conditions of an automated machine at a manufacturing plant. Programmers should therefore generally be included in the manufacturing sector for purposes of productivity analysis. It is basically wrong to include them in as "third-sector industries" just because their job titles include the word "soft."

These distinctions are not absolutely critical but should serve to make you aware that superficial arguments occupy much of the world's time. You can turn your attention back to more fundamental problems. At this juncture I would like to turn to the question of whether Japanese manufacturing industries will follow a similar path of hollowing. I'd like to approach this problem through a number of apparently unrelated discussions.

The first of these concerns Great Britain. British manufacturing industries are also hurting today but for reasons different than the U.S. situation. The main reason for their problems lies in the appreciation of the pound that followed the development of the North Sea oil fields. This appreciation has caused a loss in the ability of the British manufacturing industries to compete with foreign firms. Traditionally, British labor unions have prevented reforms in manufacturing in the U.K. Productivity has dropped and today the average income in the United Kingdom is considerably lower than that in Japan.

In the midst of this dangerous situation, Prime Minister Thatcher's Conservative government has pressed full steam ahead with a number of steps: privatization of nationalized industries; liberalization of investment restrictions; and a curbing of union power. The benefits of these efforts have yet to be fully realized. From the perspective of improvements in industrial productivity, however, they are certainly promising.

London was once the main locus of the world's money game, but Great Britain cannot rely on this alone to prop up its economy. Its new policies give greater weight to manufacturing industries; Japanese firms have responded to this by moving into Britain in greater numbers. The success of these ventures should be more widely publicized.

The Japanese approach here seems to have worked out somewhat more smoothly, but there are reasons to worry. In Japan there is a term "debtless company." Such corporations are on the rise. There is no question that this has been an ideal for a nation that struggled with a debt-ridden economy after World War II. This concern has resulted in the emergence of a smattering of corporations whose profits from capital surplus operations have swelled operating profits to a degree that cannot be ignored. If these profits are not handled with skill, there is nothing to prevent us from repeating the same mistakes made by American management.

Japanese companies have gone through every imaginable torture to reduce the costs at their manufacturing plants by several percentage points; after great pain they have finally succeeded. But when the profits realized from these painful steps are eclipsed by the profits realized from surplus capital operations, there is a real possibility that those people active in manufacturing will begin to lose their will to work.

Fortunately, Japan has yet to reach this point. After observing the fall of the American machine tool industry I would like to caution Japanese management to never forget that our economic growth had its origins in manufacturing. No decisions should be made without this point firmly in mind.

The Japanese Economy Will Not Go Soft

The question that plagues us today is what we need to do to preserve and improve our performance in the face of the harsh competition of today's world. Some people have recently begun to speak of the "softening" of the Japanese economy. The population of the third-sector industries is growing, but we should also take note of the fact that the second-sector (manufacturing) industries have begun to increase from a former state of virtual stagnation. The softening process that has followed the shrinking of second-sector industries in the United States has not been as evident in Japan. This is apparent in Figure 12.

There are some who nonetheless hold the opinion that Japan, too, will eventually see an expansion of its information and service industries, with a corresponding hollowing of the second-sector industries. However, these two sectors are not antithetical to one another. A rise in one does not necessarily cause a decline in the other. For example, the machine tool industry in the United States has suffered severely as a result of the money game, and not because it has been somehow eaten away by service industries. Indeed, good design, itself a "soft" aspect of a product, will make it better than other goods of its type. In short, the role of the soft aspects of the product has increased. This has been true for years, and there is nothing strange about it.

The importance of the industrial designer has been recognized in Japan for more than 30 years. The "G Mark" awarded by the Ministry of International Trade and Industry reflects this history. The value added by the industrial designer is included as part of the base of manufacturing industries; to say that this reflects a more important role for the third-sector industries is somewhat farfetched. That is the opinion of people who have no idea what goes on in production.

The current appreciation of the yen has led me to rediscover the real nature of money not backed by manufacturing activities — it is truly no better than wastepaper.

Japanese firms have not always done well with surplus capital. They bought American bonds when interest rates were high, only to see the dollar plummet from ¥240 to ¥150 in October 1985. A lot of firms were stung by that development.

Firms that concentrated on exporting goods have not had to shoulder currency exchange losses to that extent because approximately 40 percent of the exports were yen-based. Even so, appreciation of the yen will result in higher prices and will doubtless take its toll on these exporting companies in the future. Fortunately, the prices of products for which Japan is the sole supplier can be raised in accordance with the yen appreciation.

		Persons Employed	First Sector	Second Sector	Third Sector
Japan	1950	Unit: 1,000 35,626	48.3%	21.9%	29.8%
	1960	43,716	32.6	29.2	38.2
	1970	52,042	19.3	33.9	46.8
	1984	57,660	8.9	34.2	56.9
United States	1970	78,627	4.5	33.2	62.3
	1983	100,834	3.5	26.8	69.7
West Germany		25,173	5.4	41.6	53.0
France**		20,867	8.1	32.9	59.0
United Kingdom		21,378	1.6	32.1	66.3
Italy**		20,704	12.3	35.8	51.9

* : Definitions of industrial sectors differ from country to country
** : 1983 figures
Source: Bank of Japan, *Comparative International Statistics*

Figure 12. Economic Composition of Japan and the West (1984)*

Similarly, the costs of the materials to make these products have all dropped considerably. Just the other day, for example, when I visited a chemical fiber manufacturer in Osaka, I was told in confidence that the current drop in naphtha prices alone would result in profits for the firm of some ¥ 20 billion.

The answer to concern about "softening," therefore, is that Japan will have no problems at all as long as it is able to continue to import raw materials and add economic value through the processing of these materials. Herein lies the strength of manufacturing industries. The money game is much riskier. Changes in the exchange rate are passed on as straight losses and there can be no negotiations for price increases.

Japan Is Rewriting Economics

Let's take a closer look at Japan's success. Why has Japan triumphed as a manufacturing power? The example of Singapore, a nation one-fortieth the size of Japan, is instructive.

Singapore has a population of 2.4 million people and occupies a land area no larger than Awajishima, a small island between Osaka and Shikoku. The income level in this country is second in Asia, following only that of Japan. This tiny island has neither natural resources nor an energy supply and it has reached its current status through what might be called "trading in processing" — turning raw materials into processed goods.

But U.S. protectionist moves threaten to trample the gains that Singapore has worked so hard to attain. In the hysteria of the autumn session of the 100th Congress, Prime Minister Kuan Yew Lee of Singapore made an edifying speech. His message, "You appear to be trying to attack our very right to survive!" went into homes throughout America via television and managed to bring some of the more zealous members of Congress back to their senses.

His appeal can be paraphrased as follows. There are basically three possible ways for this small, densely populated island to continue to exist: "We can play the money game, we can rely on slow but sure manufacturing activities, or, if these two fail, we can rely on the tourist trade. The final choice is, of course, up to us, but I am convinced that the only option really open to a small island nation with the population of Singapore is to trade in processing."

Prime Minister Lee's insights are as applicable to Japan as they are to Singapore. Japan occupies a mere .3 percent of the world's land mass. What, then, has allowed it to rank first in the world in the number of motor vehicles produced each year? Japan is remotely situated. It is isolated from other advanced nations. If the world leader in auto production were a country with a land mass the size of the United States, that would make sense. I think that it behooves us to give considerable thought to the reasons that this position has been seized by a country as small as Japan. In essence, Japan is rewriting the classical laws of economics.

Per capita income in Japan now exceeds $10,000. When we consider that fewer than 20 countries have reached this mark, it is obvious that Japan packs considerable economic wallop. The average per family income is now ¥5.2 million ($36,400), and personal savings average ¥6.8 million ($47,600) per family. This wealth was generated by the productivity of Japanese manufacturing industries — productivity that has enabled the country to make outstanding products at the lowest prices. I'd like to take a somewhat detailed look at what it is that goes on in Japanese manufacturing that has provided Japan with the base for its economic power. "Manufacturing" is not something that can be accomplished simply by producing a set of blueprints. You don't just buy an automated machine and flip the switch.

The high productivity of the Japanese manufacturing industries has drawn to Japan group after group of overseas observers who wish to emulate our success. But there are things that prove remarkably difficult to comprehend if all one does is look at factories.

First, let us consider the manufacturing facilities themselves. In both the automaking and the semiconductor fields, there's virtually no difference in the equipment used in Japan and the West. It's said that the Japanese use more robots, but these days nearly every country in the world is moving at breakneck speed to install more robots.

Worker diligence is also similar. The Japanese do not work that much harder than people in other countries. There was a time when people said that American workers were not doing a good job making cars because they smoked cigarettes while they worked; you will no longer see such workers. Everyone works hard now.

The technical playing field, too, is about level. Here the Japanese are not particularly superior to people of other nations. Indeed, there are more technical people abroad with unique ideas than there are in Japan. Finally, there are no longer big differences in products. Cars, semiconductor devices, machine tools — all look about the same from country to country. Why is it, then, that Japan is so clearly leading the pack when it comes to quality, cost, and delivery time? This is not something you can understand by merely looking at a factory.

If products could be made by simply handing over the blueprints, preparing the equipment and providing job procedures for the workers, manufacturing would be a piece of cake. No matter how well conceived, a blueprint is still the product of a human being — and thus subject to error when executed. Technology changes rapidly and it is a major task just to keep the average worker abreast of new developments. There are also the external factors such as the parts and goods ordered from outside suppliers; one cannot assume that an order alone will guarantee that the parts will arrive on time and without defects.

Successful manufacturing requires careful orchestration. Each person must pay careful attention to the myriad details that occupy his or her own areas of responsibility; each person must make unceasing efforts to implement solutions to common problems. Victory cannot be achieved until every employee devotes the full measure of ability and knowledge to ensuring that the company meets its goals. This is why quality control in Japan cannot be successful without a *total quality control* (TQC) program.

In addition to having thoroughly integrated quality control programs, successful manufacturing companies are flexible. Consumer tastes change frequently and are ever more sophisticated. Popular products from one's competitors can quickly erode the competitiveness of a popular product line. If you cannot quickly develop new products in response to these factors you will lose your position in the market. These are the realities of the Japanese situation. The manufacturers who have survived what can be called the toughest competition in the world have been finely honed by the competition; their products are accepted by the rest of the world simply because they are Japanese.

As long as foreign firms are unable to find some means of opposing the Japanese way of doing things, they will not be able to penetrate the Japanese market, and trade friction will not go away. Even if we go out of our way to buy imported goods, the products will have to meet Japanese needs. If they expect us to buy exactly what they have to offer, without making design changes, they'll have another thing coming. The hard sell just doesn't work anymore.

"Principles of Fitting"

As noted above, the success of manufacturing depends upon human factors. The Japanese have long realized this and have recently applied "principles of fitting" to their overseas enterprises. One thing made clear from the experiences of many of these firms is that control systems used for overseas plants will not run smoothly unless they are adapted to local conditions. This is like the tailor who fits a customer for a new suit of clothes. The expertise developed in Japan serves as a model to be brought into line with local realities.

This fitting process works as follows. The basic pattern must be established by bringing in a system that has already been honed into shape in a Japanese factory. It is almost impossible to start from scratch on a site and put a new system together. The idea that only the brain work has to be done in Japan is the empty theory of people who have had no experience with actually making things. Expertise does not develop unless you have actually had a manufacturing facility of your own. I think it's important to realize, in this day and age, that any firm which wishes to succeed overseas must follow the above pattern.

This is particularly apparent when we study the cases in which the reverse has been true. Many Japanese firms have gotten into serious trouble by simply acquiring a going concern abroad. I have heard stories about Japanese firms who bought semiconductor plants in California's Silicon Valley only to find that it took them two years just to get employees to take off their shoes when they entered a clean room. That kind of situation symbolizes the risks extremely well.

These sorts of gruesome manufacturing experiences have increased as more Japanese companies have moved overseas. Because of this, Japanese firms have decided to build their foreign plants in rural areas. People in farming communities have always been hard workers and know from personal experience that you can't make anything if you don't work. It is in these locations that one can often start from scratch, bring in Japanese methods, and tailor them to fit the locality.

Let me give you an example. A Japanese firm that we'll call "Widgets Japan" built a color television manufacturing plant in Mexico five years ago. I visited this plant two years ago, and it was a fine factory indeed. The female employees, in particular, worked hard, and plant morale was, if anything, greater than what you would find in a Japanese factory. This was my first trip to Mexico, and I was surprised since the only image I had of Mexicans was based on Mexicans in the United States. Prior to my trip to Mexico, I had visited an American factory in California's Silicon Valley. At that facility I had found that all of the plant's expensive measuring instruments were chained down and locked. According to the person who showed me around that plant, "It's like we're working in the midst of thieves."

Due to the strength of this image, I was extremely surprised at the atmosphere in the Japanese-run plant in Mexico. When I asked about it, I found out that this plant had been built in the middle of a plain in a very isolated part of the country. The plant managers told me they had recruited their work force from the local farming community and had trained these people thoroughly. Their new employees were untainted by bad work habits. It ran just like a plant in Japan: The workers held morning meetings with calisthenics and pep talks; they sang the company song; and they went to work. The absentee rate was only about 4 percent. This, too, made this factory absolutely no different from a Japanese plant.

Finally, because Mexicans are better with their hands than Japanese, I was told, the defect rate was actually slightly lower than the normal Japanese rates. Even after hearing all this, I still had my earlier impression strongly in mind. I asked if there were any thieves around. My guide looked surprised at this question, and replied, "We get by with just three watchmen. Pilferage? I've never heard of any."

The possibility of success is very high for firms that follow this "clean slate" method. They can do well by building Japanese-style factories in places unsullied by traditional Western-style factory management, then skillfully altering the methods in accordance with local realities. But I want to stress that this cannot be done without first developing a finely honed expertise in the details of plant management in Japan.

Japanese plant management centers around the employee. Products are ultimately made by human beings and success is determined by how individual employees think and act. It's the one-by-one accumulation and utilization of seemingly minor techniques that finally leads to success. The machines that do nothing more than perform the manufacturing processes are secondary. This is not exactly logical in Western terms but it is truly the secret of our success. For example, it is often the case in factories built by Japanese firms that plant managers will eat their meals in the plant cafeteria alongside the workers. This practice is almost never seen in Western factories, and observers are inevitably surprised by it.

I'm sure you understand what I'm getting at with all these examples. It is a major mistake to think that one can manufacture good products at low prices by merely relying on a logic that exists only in the head.

It is not my intention here to do nothing but recite tales in praise of Japan. The object of these tales is to impart an understanding of what it means to engage in manufacturing activities in today's world and, in particular, to give you an idea of what is involved in recent developments. The real roots of trade friction lie in production issues, not in political posturing. An understanding of what *really* goes on in manufacturing will make it easier to come up with policies that will help solve the problem.

Production at manufacturing plants could be termed a "hard" process as opposed to the "soft" processes of computer software and systems technology. However, hard processes are planned on the basis of the soft processes.

This is a key element in production efficiency and productivity. The special attractions of Japanese methods are that they are able to adroitly link human beings with their manufacturing equipment. You could, if you were of a mind to, call this soft, but no factory could function without this type of process. In Japan this is more correctly called "control technology," and it is the major factor in differences between U.S. and Japanese productivity.

What is control technology? In more simple terms, it amounts to about the same thing as the technology needed to drive a car. When operated by a bad driver a car can be a lethal weapon. In other words, a car serves no useful purpose without the process we might call driving technology.

Control technology facilitates the manufacture of value-added products. How this has been done has varied in the several developmental stages of the Japanese economy as well as by industry sector. A short review should be most instructive both as history and prognostication.

The Power of the Materials Industries

In the 1950s and 60s the impetus for Japan's high level of economic growth came from the steel, shipbuilding, petrochemical, and fabric industries; the so-called heavy chemical industries stood at the center of this growth. This period was followed in the 1970s by the emergence of assembly industries, starting with televisions and expanding into automobiles. The energy crises of 1973 and 1978 came as a severe jolt to those industries since they handled difficult-to-machine materials. The rapid introduction of technologies, however, permitted sweeping reductions in energy usage. Also, new machining methods allowed most firms to maintain their competitive edge. In fact,

Japan's GNP has increased by two-and-a-half times since the second fuel crisis, while the petroleum imports dropped by 20 percent.

Productivity gains also extend beyond the assembly industries. There are those who say that sectors such as the materials industries, which have low levels of value added, are no longer any good for Japan. In spite of these claims, however, certain of these industries have come along steadily in recent years. Often their progress takes place outside the public eye.

I'd like to draw your attention to Figure 13. This table is based on research I did in 1980 during a project on the fifth generation computer and may be somewhat out of date by now. I'd like to include it nonetheless. It is a portion of a list of some 70 different corporations that have each captured close to 50 percent of world demand for their products. The numbers may have changed a bit, but the truth remains that many of these firms are located in Japan.

Although there are exceptions, nearly all of these firms are of small to medium size. They share several features:

- very few of them are famous;
- they are not based in Tokyo or Osaka but are what might be called regional companies;
- they specialize in parts or materials rather than assembly;
- and they don't maintain a very high profile in the market.

For these reasons, perhaps, the bigger companies have shown little interest in trying to steal their business.

Most important of all, the products that these companies provide are parts and materials that their buyers simply cannot do without. Therefore, they typically have close relationships with the Japanese automobile and electronics industries, among others. We could go so far as to say that the very presence of these products and firms in the background has allowed the major Japanese industries to gain the competitive power that lies behind the current trade friction!

Corporation	Product	World Share	Domestic Share
Yoshida Kogyo	zippers	60%	95%
Tokyo Aviation Instruments	small aircraft instrument panels	65	100
Dainippon Printing	television shadow masks	50	95
TDK	ferrite	46	85
Kyocera	IC packages	60-75	90
Mabuchi Motors	magnetic motors	70	92
Epson	small printers	60-70	90
Tannan Seisakusho	aircraft engines	70	95

Based on 1980 survey

Figure 13. International Market Share of Selected Japanese Corporations

A second factor accounts for the importance of these "background industries." They have thoroughly automated and have made virtually all the necessary equipment in-house. This has made it possible for them to move at the drop of a hat into overseas production and keeps their particular expertise and knowhow from leaking to outside competitors. In point of fact, a visit to the factories of these manufacturers will leave you with the strong impression that you have just come across a machine tool builder.

A quick look at this list should lead to an understanding of the characteristics I have just described. A good example here is Yoshida Kogyo, famous for its "YKK" brand fasteners. The company has grown quite large, but it is still known only in the fastener market. Its main plant in Kurobe resembles nothing more than the plant of a machine tool builder. The same is true for the Kyocera ceramics semiconductor packages. It is my firm belief that this list illustrates one of the paths that Japan must take for its future survival as well as some basic principles American managers must learn to ensure *theirs*.

The world's natural resources are not evenly distributed. For example, close to nine-tenths of the world supply of niobium is produced by Brazil, but people do not complain that Brazil is hogging the market because of this. The gods have not been equitable in their allocation of natural resources, and there is nothing to do but accept the fact that some things will be made only in certain restricted areas.

Japan does not have much in the way of natural resources. It has, instead, 120 million outstanding people. My way of thinking is that we should compensate for the lack of natural resources by pooling our efforts and producing *manmade resources* in their place.

The assembly industries with the strictest standards in the world are in Japan. It is my firm belief that, if we can profitably produce parts and materials that satisfy the user — and do this more reliably than any other country in the world — the world will continue to buy these parts and materials to use in their own manufactured goods.

In a sense, this will be effected by increased emphasis on *value-added* products by these industries. Shipbuilding provides a contemporary example here.

The Future Belongs to Manmade Resources

Recently it has come to be said that Korea has overtaken Japan in the shipbuilding industry. But the engines — the key part in any ship — the radar, and many other essential components are virtually all supplied by Japan. In other words, Japan supplies the most critical and value-added parts needed in shipbuilding.

The same is now true with automobiles. The Korean cars that have risen to the top among imported models have been based on a model produced by Mitsubishi; most of the engines are being supplied by Japan. Finally, it is being said by some that appreciation of the yen will cause Japanese electronic industrial equipment manufacturers to be devoured by their Taiwanese and Korean competitors whose currencies are linked to the dollar. But virtually all of the components for these products are being supplied either directly from Japan or from Japanese-run plants in Southeast Asia.

I once had a conversation on this subject with a steel manufacturer. He told me that the steel industry is continuing to hang on, too. As noted in Chapter One, Japanese steel manufacturers have succeeded in creating the high-tensile steel critical to the auto industry and are supplying it in huge quantities. Vehicles made with this steel are lighter and their fuel economy increases accordingly. All Japanese automakers now use this steel.

Overseas automakers want to use this steel too. It has become a deciding factor in the fuel economy wars. The materials used to make this high-tensile steel are not being produced abroad and as a result Japan's monopoly in this material is likely to continue for some time. Those who wish to raise the performance of their completed products will have no choice other than to bite the bullet and buy from Japan, no matter how distasteful this may seem. The highly touted Korean and Taiwanese cars would not exist without Japanese high-tensile steel.

This presents an interesting irony. If Japan were to go out and capture half of the world market for completed cars, all kinds of fuss would be made about competition. No one worries about overreliance when they are able to use Japanese steel and reduce the weight of their own cars by 10 percent. They simply clamor for the sales.

It is often said that we have entered the "information age." What we really have is the "selective buying age." Earlier I noted the problems of the American machine tool industry, but there are still American machine tool builders with a strong competitive edge. These include firms such as Cincinnati Milacron, which specializes in large machinery, and Moore Special Tools Company, which makes jig borers. Today these firms continue to dominate their markets; there has not been much Japanese firms can do about it. Their management has favored long-term stability and is in business for the long haul.

The United States is smart enough to realize this. Not long ago, a certain American economics magazine carried an article that predicted that by 1990 nearly half of the cars on American roads would have engines made in Japan. This prediction was based on a total of: cars made in Japan; cars made in the U.S. by Japanese firms; cars sold by Korean firms, by General Motors, and by Ford that were manufactured OEM in Japan; and cars made in Europe that contained Japanese engines. The article said that together these would amount to nearly 50 percent of the total U.S. engine market. The article then went on to comment: "Rolling steel and making auto bodies are jobs that can be done anywhere in the world. The parts of the car with the highest level of value added are the engine and the transmission. These are what Japan has its eye on, and it is gradually solidifying its position in these areas."

Japanese automakers have had to contend with the strictest emission standards in the world. This struggle has finally begun to pay off. Japanese-made engines have high fuel economy and are light, with high horsepower output. They run clean. And the recent tightening of emissions standards has thrown the spotlight further on Japanese engines since they meet these new requirements.

The major lesson to be learned here is that half-baked, sloppy work will never sell to today's selective buyers. Anyone who makes a product that his competitors cannot match will inevitably take half or more of the world market. This is the path Japan must take to continue to survive. There is another lesson here — we should set our sights on producing the manmade resources that manufacturers all over the world will use to make their own completed products, whether materials, components, or products such as robots. We can be satisfied if people buy these products from Japan and use them to add value to their own products, thus enriching people's lives.

Manmade resources or value-added products require trained minds in their design and creation. This is a further source of Japan's manufacturing strength. Until a short time ago, small and medium-sized firms in Japan were seen as living in poverty next to large firms. The situation has reversed today; the profitability of the small firms contained in Figure 13 is considerably higher than that of the large corporations. These days, it's the presidents of these firms who are able to buy Mercedes Benz sedans out of their pocket money. As I have mentioned, such firms are a major engine of economic growth.

Japan has been in a better position to capitalize on this state of affairs than any other country in the world due to the large numbers of engineers that our universities graduate each year. The Young Report on American competitiveness notes that there are 73,600 graduates from technical programs in Japan each year, and just 67,400 in the United States. Even though the United States has twice the population of Japan, Japan now has more engineers. At this rate, it would appear that United States manufacturing will eventually be overwhelmed by Japan.

After World War II, universities were formed throughout Japan. One critic referred to these, half scornfully, as "train lunch colleges." What he meant to imply was that any place big enough to have a train station that sold lunches now had a college.

Nonetheless, hindsight shows that these colleges have been a key factor in the strengthening of the Japanese economy to the point where we now have a GNP of $10,000 per capita. Some 96 percent of Japanese young people go to high school, and 37 percent go to college. This supplies a large number of new, fresh graduates to be pumped into Japanese companies each year. The quality of these graduates is, of course, an issue. Even so, their sheer numbers make it extremely difficult for other countries to match Japan's gains in productivity and developments. France and the United Kingdom lead Europe in the number of engineering graduates, but they turn out no more than 12,100 and 10,000 engineers a year, respectively.

If you have any doubts, all you need to do is attend an international academic conference. Not only will there be many Japanese participants, but more and more outstanding papers are being presented by them.

At such affairs, the most popular sessions are those that deal with industries in which Japan is particularly strong internationally. Semiconductors are one popular topic, of course. The semiconductor market is one characterized by extreme volatility and it is worth noting that at international conferences on the subject, as many of the important papers are presented by Japanese as by Americans. This shows an incredible level of energy.

Research and Development Fills the Gap: A First Look

Next to education, research and development investment has been a significant factor in the preeminence of Japanese manufacturing. Its role in the future will be even greater. In the past few years Japanese corporations have reached the point where they are able to pour huge amounts of money into research and development. Without money, all the engineers in the world cannot make research and development a reality.

The amount of money spent here in the last few years by Japanese corporations is staggering; it is not unusual to see firms in the electronics industry with R&D budgets as high as 10 percent of their sales. The auto and materials industries are also spending huge sums of money. In 1984, the portion of the GNP claimed by research and development in Japan was 2.8 percent; in the United States it was 1.8 percent. Japan's R&D spending on nonmilitary items, in particular, outstripped that of the United States.

Such spending must be seen as a form of long-term investment. There are people who wonder why Japan makes few revolutionary new discoveries. It should be pointed out that the above level of spending has been in effect for only the past four or five years. So I think it's safe to say that revolutionary new discoveries will materialize in the near future.

In conclusion, there are but few countries with the potential to achieve a successful balance between the broad divisions of materials and assembly industries. Japan has achieved this goal and as a nation of 100 million highly educated citizens is indeed formidable.

This population is neither too large nor too small. Some suggest that the ideal population of a small island nation such as

Japan is about 30 million; that level, however, would have made Japan's diversification and large-volume consumer market impossible. The fact that our next-door neighbor, Korea, has less than half of our population has already put that country at a disadvantage. If Taiwan had to rely on its domestic markets, it would never be capable of more than small-scale production.

Today Japan is at a crossroads. The incredible energy and productivity of its people can cut two ways. If improperly handled, its sheer size alone can have disastrous effects. We must remember that it is impossible to apply the brakes once this energy is set in motion. Poor navigation could sink this giant ship and defeat our potential before it is ever fully realized.

The complex interrelationships that make up the international economy make it clear that Japan should continue to increase its output of those manmade resources that only it is capable of delivering. This would be the country's best role and one it is already playing in the production of RAM chips. Japan is already the monopoly supplier of these critical components in computer production. The result has been gratifying to Japan, but perhaps ironic to the United States. U.S. tariffs on these semiconductor devices have forced an increase in their cost. This has driven up the costs of American computers and thrown a wrench in the works for the U.S. computer industry. Furthermore, since Japanese computer manufacturers can buy these devices at lower domestic prices, Japanese computers now have an absolute price advantage. In this case, Americans have wrung their own necks.

The RAM situation illustrates one strategy for Japan's continued survival; it is a direction in which we have already begun to progress. I'd like to propose that we actively pursue this type of policy. Japan has already reached the point where all the preliminary steps have been taken: we have large numbers of engineers graduating each year; we have a variety of assembly industries that give direction to our development and allow us to test our accomplishments; and we have the "ultimate arbiters" in some of the world's most finicky consumers. I fully expect all of these factors to work together to make Japan the world's most important supplier of manmade resources.

CHAPTER 4

Productivity: The Key to Manufacturing Success

Children always have better lives than their parents, and the third generation will have an even richer life — this way of thinking is taken as a given in the United States. But our children are probably looking at a lower quality of life than we enjoy today...

These words were used to sum up a 1980 American television show broadcast on NBC called "If Japan Can, Why Can't We?" They remain firmly lodged in my memory.

This 90-minute show pointed strongly to drops in productivity and quality in American industry. It provoked strong reactions from throughout the industrial community in the United State; it is said that more than 6,000 copies of a videotape made from it were sold in the United States.

Japan has worked ceaselessly since World War II to rebuild its economy. Through these efforts we have become the world's second most powerful economic force. But the crisis in American industry now threatens to overwhelm Japan as well.

Despite overall success, Japan's economic growth has been accompanied by a growing number of problems: the ever-increasing barrier imposed by trade friction; increasingly keen competition from the newly industrialized countries; the trend among domestic companies to move their manufacturing facilities abroad to cope with these problems; the inevitable aging of the nation; and massive government debts. Any one of these problems alone would be difficult enough to solve. If we fail now to deal decisively with the problems despite the wounds it might entail, the NBC prediction may become a part of our future, too.

Japan's 10 Percent and What It Means

It is unlikely that the Japanese economic position will ever erode, but if it does the effects will quickly spread outside the Japanese national boundaries. There are numerous examples from the past of the chaos that can ensue when a nation with 10 percent of the world's gross national product sneezes.

This was made marvelously clear in 1973 during the first energy crisis. This caused Australian iron ore exports to drop drastically. As a result, the port facilities used for iron ore shipments bound for Japan grew idle. This led a minister in that country to submit his resignation.

At a time when the Japanese economy was in its period of rapid growth, a politician in Europe cracked a bad joke. He said the world would be a pretty good place to live if Japan and the Soviet Union would just disappear. That jokester had obviously forgotten the fact that a drop in exports to Japan would have a disastrous impact on the economy of many countries in the world.

After the United States, Japan is now the second largest food importing nation in the world; it is the largest importer of marine products. If the export flow of high-valued marine products such as shrimp, crab, and tuna were to be shut off, several newly emerging countries would face bankruptcy. Furthermore, exporters of raw materials such as iron and coal would be in serious trouble if they were to lose their Japanese markets. Factories would shut down in both the advanced nations as well as the newly industrialized Asian countries.

There are people who hold the opinion that the world order is molded by the United States and the Soviet Union, and that Japan merely swims in the reservoir that these superpowers have dug out. The fact that one-tenth of the world's economic activity now comes from Japan means, to all economists, that the country must be taken seriously.

Japan's 10 percent share of the world economy has other effects. The recent appreciation in the yen has had a shock effect not only on the Japanese economy but on those countries that rely heavily on certain Japanese goods. The effects have been particularly terrible — in the form of escalating prices — on those countries for whom Japan is the *only* supplier of certain goods. Some newly industrialized nations in Asia are reaping the benefits of the stronger yen, but they are also spending more since Japan supplies 70 percent of the components used in Korean VCRs, as well as the engines for Korea's most popular export car.

Japan is also the world's leading supplier of the dies critical for high-volume mass production. You may think that anyone could make dies if he had the right equipment, but this isn't so. You need years of experience as well. Production is impossible if a die cannot be fabricated simultaneously during the design, testing, and mass production of a new product.

Japan delivers dies quickly. Japanese manufacturers use numerically controlled (NC) equipment to allow them to supply dies in one-tenth the time required for hand fabrication. To the customer who wants to get a jump on his competitors this kind of turnaround is a godsend. People are happy to buy Japanese dies even if they are more expensive than others.

Even given the essential nature of some Japanese products, there are still limits to the price increases that foreign companies can absorb. In time, these increases will drive up the prices of goods from the newly industrialized countries.

The same economic rules apply to the United States, the country that ignited the flames of the higher yen in the first place. As I have mentioned, Japan is rapidly becoming the sole supplier of certain types of memory chips needed for computer production. This provoked dumping charges, and the price of the Japanese chips more than doubled. It is predicted that the U.S. trade deficits with Japan will increase and ultimately reach $800 million in 1986. Japan is not alone in singing the blues about the stronger yen.

We now hear some people say the only real solution to the present trade imbalance is for Japan to become a consumer superpower. There is nothing particularly surprising about saying that Japan works too hard and should relax a little by spending money. But no country can be a consumer superpower without first becoming and staying a production superpower. A country with low production levels cannot sustain high consumption levels indefinitely. On the contrary, every country in the world should strive to obtain high productivity levels, and to use such productivity to sustain high levels of consumption.

It is the difference in productivity that most graphically accounts for the difference in competiveness between the United States and Japan. This was not always so. My first trip to the United States was in 1958. At that time America's industrial might was overwhelming. Wage levels in the United States were five times higher than Japan. But the Americans didn't seem to be working expecially hard. The products they made were far superior to those made by Japan; their costs were from 20 to 30 percent lower, depending on the individual product. That was at a time when the idea of a two-day weekend would have been unthinkable in Japan. Productivity in U.S. manufacturing industries was up to four times higher than in Japan. In short, I learned there that productivity is the true source of economic wealth.

Today, 20 years later, the relative positions of the two countries have reversed. The United States has, of its own accord and by itself, caused the hollowing of its manufacturing industries. It has subsequently lost its competitive power and begun to decline.

The U.S. insistence that Japan too should sink is a root cause of trade friction that does no one any good. If this were to take place, our grandchildren will be forced into lives poorer than ours, just as the NBC broadcast predicted. This is why it is so important to acknowledge that productivity is the starting point of the wealth that has become a part of our lives today.

High Productivity Leads to Wealth

Productivity is not a new concept by any means. It has played a role since the beginning of civilization. The people of the Stone Age spent their entire day scrounging for food. The invention of agriculture dramatically improved productivity in the production of food and gave people time for other pursuits. Civilization and all its trappings developed from this phenomenon. The Mesopotamian civilization born by the Tigris and Euphrates, the Chinese civilization that developed from the Yellow River Basin — both these and indeed all civilizations date from the discovery of agricultural technology.

Put in other terms, productivity is a measure of the value added by the labor of human beings. This productivity gives birth to wealth and the basic principle remains true today.

It is possible to trace the competitive power of today's corporations through the yardstick of productivity. Some 60 cars a year are produced per worker at the factories of the Japanese automakers; the American average is 29 vehicles. "Competition" is hardly an issue when productivity differs by this extent. This kind of productivity has increased the Japanese share of the world market to the point where trade friction is almost inevitable. Surplus value created by this high productivity has enabled corporations to invest in research and development and upgrade their equipment, thus further increasing their competitive edge.

The term "productivity" is so often misunderstood that I would like to clarify it, at the risk of boring my readers. The misunderstanding revolves around the notion that productivity implies the intensification of labor efforts and leads to increased unemployment. I was surprised and shocked to hear this misconception repeated by a high union official at an OECD symposium on productivity held in France in the fall of 1985.

My response to him was as follows: "There is a fixed definition of productivity. Productivity is efficiency. Some may say that lengthening the amount of time spent at work will result in more products being made, and that this is equivalent to an increase in productivity; it is nothing of the sort but the opposite.

"Let us suppose that someone has devised a new way of doing a particular job, so that what once took 10 hours now takes but five. The worker would finish the job sooner and would feel more rested. Now, for the first time, we can say that productivity has doubled. Thus, increases in productivity are actually acts of human respect. It won't do to say that this wasn't the intention. That's the definition. It's not true to say that an increase in productivity will result in someone getting fired.

"Many people say that there are manufacturers in Japan that have the highest productivity rates in the world. And note that it is also true that our unemployment has been in the 2 percent range for more than 20 years, and this is the lowest in the world. In point of fact, nations with low productivity in the manufacturing industries have high unemployment rates."

Figure 14 shows this point clearly.

I further corrected the man's misconceptions by speaking about robots: "Some people believe that the introduction of industrial robots will lead to more unemployment; in reality, the opposite is true. Statistics show that Japan has 70 percent of the robots working in the world. I've never heard anything about people getting fired in Japan because of the use of robots. In fact, our unemployment rate is the lowest of any of the advanced nations."

It would be more accurate to say that the use of robots has allowed Japanese manufacturing industries to increase productivity and maintain our international competitive edge. In turn this has guaranteed employment stability. If, for example, Japan had been late in its use of robots, it would have lost out in international competition. This would have meant increased unemployment.

Then I showed them Figure 16, showing the trends for the prices of manufactured goods and services in Japan over the last decade. As this chart shows, price increase rates have varied considerably during that period; in some cases prices have remained the same, or even dropped.

If you look into this a bit, you will find that the reasons are simple enough. The prices for goods that have benefited from

Year	Japan	U.S.	EC	West Germany	France	U.K.
1979	2.1	5.8	5.5	3.8	6.0	5.1
1980	2.0	7.0	6.1	3.8	6.4	6.4
1981	2.2	7.5	7.8	5.5	7.8	9.9
1982	2.4	9.5	9.4	7.5	8.8	11.5
1983	2.6	9.5	10.6	9.1	9.0	12.3
1984	2.7	7.4	10.9	9.1	10.1	12.7
1985 (first half)	2.6	7.2	10.6*	9.4	10.6	13.1

* : EC figures for 1985 include 10 countries, including Greece; all other years
include 9 countries. Figures for 1985 have been seasonally adjusted.
Source: Japanese Ministry of Labor, *Labor Statistics Survey Monthly Report*
(July 1985)

Figure 14. Unemployment Rates of Major Countries

Japan	41,265	Czechoslovakia	1,845
U.S.	9,400	U.K.	1,753
West Germany	4,800	Canada	700
France	2,010	Australia	528
Italy	2,000	Belgium	514
Sweden	1,900		

Source: Robot Institute of America

Figure 15. Robots in Operation (as of December 1983)

Foods	'85/'76	Manufactured Goods	'85/'76	Services	'85/'76
Rice (Grade B) (10 kg)	144.1	Electric washing machines	91.8	Rent in privately owned dwellings	142.5
Raw tuna (100 g)	139.5	Color televisions	88.0	Carpentry work per hour	153.5
White radish (daikon) (1 kg)	168.8	Cameras	86.8	High school tuition	344.4
Onions (1 kg)	87.5	Passenger cars	119.4	Haircuts	154.4
Pork (100 g)	87.5	Gasoline	118.9	National railway tickets	466.7
Eggs (1 kg)	102.8	Pantyhose	123.5	Postcard postage	200.0

Source: General Affairs Agency, Statistics Bureau, *Monthly Price Report, Retail Price Section* (August 1985)

Figure 16. Price Increases (Percent) — Tokyo (1976-1985)

greater productivity in the manufacturing process over the past ten years have either increased only slightly or have dropped. This is even true of food, not strictly a manufactured product. Many people are under the impression that food prices in Japan are constantly rising. That is not entirely true — in some cases they have actually dropped.

Wages, too, have increased by 220 percent in Japan during the past 10 years. What has happened, in short, is that Japan has worked like a nation possessed to increase productivity; these frantic efforts have given new wealth to the Japanese people. Furthermore, they are the source of a powerful new world economic position.

This is only natural, but I'd like to explain in a bit more detail why productivity is the most fundamental element in our search for routes to our future survival.

Years ago I served as the chairman of the Tokyo Prefectural Office Automation Committee. The Committee was charged with planning automation of government offices. I received the following question during a committee meeting: "The automation of more offices in Tokyo will result in there being too many people. Won't they start getting fired?"

I answered, "There will be too many people if offices are automated. If office automation didn't cut the number of people on the staff, it would be better not to do it. But it doesn't necessarily follow that people will be out of work just because there are too many people. Tokyo has a large number of public facilities. There are halls for the aged, libraries, sports centers, and lots more — their monetary value would be incredible. But what's being done with them? They all shut down at five in the evening. And they're shut on holidays, too, which is a perfect time to use them.

"I don't care how nice a library might be; once you've shut the door, it's not a library to the people any more. It's just an empty concrete shell. And if you ask why these facilities are shut so often, you always get the same answer — because there aren't enough people available to work in them.

"So why don't we implement an office automation program, and use some of the extra people to keep these offices open at night? In other words, office automation is almost a prerequisite for expanding our social services. We used a lot of tax money to construct these facilities, and I think it's a real waste to restrict their operating hours just because there aren't enough people to work in them."

This proposal was accepted immediately, and the Tokyo Prefecture decided to keep libraries open through the evening. After the program was implemented a survey showed that usage had nearly doubled.

This type of thinking is important. In brief, by raising productivity we get more economic space to play with; we can then use this new space to start up something that we have never been able to do before. If you visit the advanced Western countries, you will see that they have a large number of public facilities, most of which are free. Some observers have criticized Japan for lagging behind in this respect. Very few, however, stop to ask themselves where the societies got the money to provide these facilities in the first place.

Those who proclaim what course Japan must take in the future must achieve a correct understanding of the source of Japan's wealth. If they do not, their ideas will have no more substance than a picture of a delicious meal or an empty theory someone thought up from behind a desk. Before Japan is able to become a consumer nation, we must first decide how we will produce the money for all this consumption. The answer, of course, is to further strengthen productivity; without this decision, everything else will be meaningless. Making money is not as easy as some might believe.

The international market today is highly competitive. Any company that backs off or sits down on the job for even the shortest time will wind up in bankruptcy. Unemployment has risen where people waste their time in worthless arguments — such as saying that the introduction of robots will result in people losing their jobs. This is scarcely an exaggeration but something that people will eventually come to understand. Even those countries that give priority to the common worker as a matter of political philosophy have begun installing robots in place of

A Japanese automobile factory using industrial robots

human workers. This has occurred in Eastern Europe. There is no place for philosophical argument in the business of strengthening your competitive power.

Japan Is Already a Magnificent Consumer Superpower

Recently some critics have called for an increase in domestic demand in Japan as a way of sopping up productivity. Japan's trade surplus, it is said, should be cut by spending more money at home. They hold essentially that the Japanese are very good at making money, but fail to spend enough of what they earn. On the contrary, if the Japanese have enough money in their possession they do a remarkably good job of using it.

The pattern has historical antecedent. You can find good examples of it in the merchant culture of the Edo period (1600-1864). The government of that period took the attitude that "luxury is our worst enemy," and tried to foist this maxim on the merchants. Even so, they managed to create a splendid consumer culture.

The Japanese also spend money on travel abroad. Japan's high level of economic growth has given rise to a major increase in the number of people who take foreign trips. At the beginning of this growth period, anyone who heard the expression "go abroad" immediately made insulting remarks about agricultural cooperatives. That was what you might have expected at that time, since the Japanese food import control system greatly enriched the nation's farm communities. During their slack season the farmers abandoned their ideas concerning leisure and began setting out on trips abroad in ever-increasing numbers (see Figure 17).

The farmers were unable to speak any foreign languages, so they trooped around the world together, following a guide and a flag. Some people scorned this type of activity; to a white collar worker of that time, a trip abroad was as unattainable as a flower blooming on a mountain top. I think it was a wonderful thing that these farmers could travel abroad. People living on the land, who since the Edo period hadn't even so much as laid eyes on the ocean, now had the opportunity to travel across it and see the world.

Narita Airport on the day before Golden Week

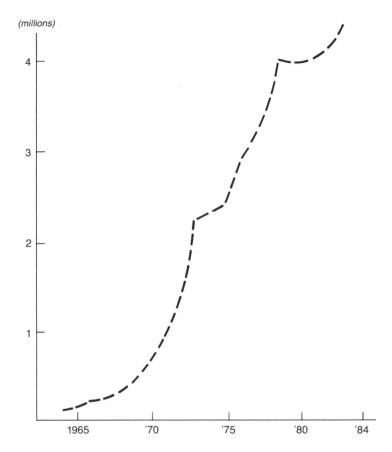

Source: Ministry of Transport

Figure 17. Japanese Visitors to the United States

The Japanese also spend their money on domestic travel. Traditional rustic inns at hot springs that had been forgotten for some time are now filled with a boisterous young clientele. Small groups of middle-aged women travelers now stand out in the first-class cars on the nation's trains. Lured by the slogan "have a nice middle age," they are walking all over Japan.

Domestic consumption of foreign luxury goods has also risen. These days, for example, women will almost inevitably have a brand-name handbag from Italy or France. These bags are so popular, in fact, that fakes are now appearing in large numbers. You couldn't expect them to appear in a market where there was no demand.

I wish those who keep saying that the Japanese should buy more foreign products would bother to study the question — these days any Japanese who is not using a number of foreign products in his or her daily life would be a rare sight indeed. To test this theory out, all you need to do is to ride a crowded commuter train. You will see brand-name foreign handbags, neckties, and accessories. It may be hard to tell with just a brief look, but you will also find women dressed in British clothes, French scarves, and Italian shoes — there's no end to it. When diamond prices fell recently, foreigners rushed to buy them. Nothing happened in Japan since diamonds had been selling well all along. Japan has become the world's best market for diamonds.

Domestic spending, furthermore, is not limited to luxuries. There will be people who try to deny this, saying that Japan's social investments are still weak and that the average home in Japan resembles nothing more than a rabbit hutch. But the rate of home ownership is now the second highest in the world (see Figure 18) and the number of residences is now increasing at a faster clip than the number of families. The average Japanese can now live on his or her own home, small though it may be.

Houses are also getting larger. With the exception of those in the large cities such as Tokyo and Osaka, the average surface area occupied by newly constructed houses in Japan now ranks among the top five civilized nations in the world. Roads are also getting better. A number of rural freeways now exist where one can drive along without seeing many cars. Japanese infrastructure is highly praised abroad for its good design and construction.

Finally, there has been significant investment in health and education. The health of the Japanese is among the best in the world. Longevity and infant mortality rates are the best among the advanced nations. Educational standards rank alongside those of the United States as the best in the world. The education of the working class is especially good.

	Japan	United States	West Germany	France	England	Italy	Sweden
Residential dwellings per 1,000 population	324 (1983)	398 ('81)	418 ('81)	437 ('82)	388 ('81)	321 ('77)	441 ('82)
Newly constructed residential dwellings per 1,000 population	10.0 ('82)	4.4 ('82)	5.9 ('81)	7.2 ('81)	3.3 ('82)	2.7 ('78)	5.4 ('82)
Average rooms per dwelling	4.8 ('82)	5.1 ('79)	4.7 ('81)	3.9 ('81)	4.4 ('82)	4.1 ('77)	4.7 ('82)
Floor space for average new dwelling (sq. meters)	93.9 ('82)	136.1 ('80)	98.8 ('81)	77.1 ('74)	—	—	103.2 ('82)
Percentage of home ownership	62.3 ('83)	65.6 ('80)	37.5 ('78)	50.7 ('82)	59.0 ('82)	50.9 ('71)	38.9 ('75)

Source: Ministry of Construction, *Construction Statistics Overview*; United Nations, *Annual Statistics*

Figure 18. Housing Standards

I could give many more examples about the quality of life in Japan. Anything made by human beings probably falls short of perfection. In terms of averages, however, there is no doubt that we have reached a high enough position to be the focus of envy from people in other countries.

The major characteristic of modern industry is mass production. The size of the Japanese population ensures us that there will be sufficient fundamental demand to support domestic production. From the perspective of the businesspeople of some of the smaller countries of Europe, Japan's market must seem like a dream come true. That is exactly why they are always demanding that the Japanese buy more.

As long as Japan remains committed to the economic principles of free competition, however, it won't be such a simple matter. Japan has no time zones, and the sun rises and sets at virtually the same time throughout the nation. Since the information media are all highly developed everyone shares the same information. So if a given product is good, it will sell throughout the country, but if it is no good, people won't buy it. A company unable to win under these harsh competitive conditions won't be able to do a dime's worth of business.

Japan's wealth is a result of its vigorous economic activities. Since we are not blessed with natural resources, we must import them from many countries throughout the world, process them, and preserve our economy on the basis of the value we add to raw materials.

Japan is ringed by the ocean. The cost of importing iron ore from Australia is a mere ¥1,500 per ton. Japan can thus buy its natural resources from the most profitable places in the world, process them, and export the finished products. The cost of shipping a car to the West Coast is in the neighborhood of ¥50,000 yen. The presence of the ocean, in other words, has the benefit of making Japan close neighbors with those who supply natural resources and those who buy our products.

The United States, too, is ringed by oceans; this fact, however, is not always beneficial. The steel industry in the United States developed in the Great Lakes region. Iron ore, limestone, coal, and the other raw materials needed for steelmaking are all found together there. Although this is the kind of proximity the Japanese could envy, the iron ore is not up to the quality Japan currently demands. Because Japan can acquire and process the best iron ore and coal in the world more cheaply, the United States is at a disadvantage. It's still out of the question, however, for the United States to import foreign iron ore all the way to Pittsburgh. Japan's success with its coastal foundries has been a major cause of the decline of the steel industry in the United States.

Until a very short time ago Japan's lack of natural resources was very much a negative factor. Further, the need to export finished goods to far off markets was also a decisively unprofitable thing to do. Japanese intelligence and technology have together worked to turn these handicaps into positive virtues. The creativity of the Japanese people has often been disparaged, but the type of intelligence shown in this case is not found in continental nations such as the United States or Europe. It's completely native to Japan, and it's what has allowed us to succeed.

Nevertheless, there are countries that are hot on Japan's heels. When they catch us, as anyone can see, we will have to increase our expertise at adding value yet further. Without this step we will not be able to maintain our prosperity. By the year 1990, Japan will be forced to effect another major transformation in order to maintain its position in the world economy. Productivity as the basic principle that will allow this to happen. There are some basic policies we must implement in order to maintain our position as a production superpower. The recent appreciation of the yen has made the situation even more urgent. In the following section I have tried to address these issues.

The Appreciation of the Yen and Japanese Staying Power

There are many people who see the recent appreciation of the yen as a major problem for an exporting nation such as Japan. The United States has adopted the simplistic attitude that if prices of Japanese imports are increased, these imports will fall in proportion to the amount of the price increases; for this reason it induced the yen appreciation. On the other hand, though, the appreciation of the yen has resulted in lower prices for all the raw materials that Japan imports. Japan's very life-blood depends on the added value derived from processing these materials. If this added value were still eroded by the stronger yen, that would be another story indeed. As long as the added value does not drop off, however, the stronger yen shouldn't even itch, let alone hurt.

In this respect we are completely different from countries that export natural resources. When oil prices fell, the amount of the drop came straight out of the revenues of the oil producing nations, with terrible results.

Even in the United States there has been talk about "saving" the oil industry. Some 6.3 percent of American industry is linked in one way or another to oil, so reverberations are felt through the entire economy if oil prices drop sharply. In the Soviet Union, where 40 percent of external revenues come from oil, these sharp price drops spell disaster. The same can be said of China. Ironically, this shows us just how fortunate Japan is to have so few sources of natural resources. Of course, problems can arise before price adjustments have been completed. Once stabilization has been reached, however, Japan will have no problems in maintaining its former position as a major economic force.

What Japanese corporations must do is devise policies that will get them over the hump during transitional periods. This is, of course, a very difficult task. No one ever dreamed that the yen would appreciate by as much as 40 percent. No matter how skillfully a firm might be managed, this is not a problem that can be dealt with simply; it's a frantic situation.

There have been critical remarks to the effect that price increases abroad of Japanese products have not matched the pace at which the yen has appreciated. When viewed from the perspectives of Japanese corporations, this is only natural.

Any company that increased its prices by the amount the yen has appreciated would be driven from the market and would, in effect, commit corporate suicide. Price increases would disrupt the sales networks that Japanese companies have worked so hard to establish go by the boards now, and mending these networks would cost more in terms of both work-hours and time than it would to rebuild them from scratch. Once you have lost credibility, it's amazingly difficult to get it back.

When you get down to it, a sales network is a form of production equipment. Firms are gritting their corporate teeth and working as hard as they can to maintain their networks. This kind of thinking cannot be comprehended by someone who spends the entire workday riding a desk. Such people should consider the unenviable position of those who are forced to maintain their market positions even though the more they sell the more they lose.

Thus firms are striving to hold their price increases to a minimum to give themselves time to find ways to cut costs and supplement their revenues. To be defeated in this struggle means corporate death. We still don't know how long this will take and corporations without the strength needed for the struggle will doubtless fall by the wayside during the fight.

There are, though, some industries that are making out like bandits in spite of the run-up of the yen. These are companies that have captured a virtual stranglehold on the world market for their products. Products that *must* be bought from Japan, in short, can be increased in price to virtually any level without fear. There are many such products in the electronics field and in heavy industry. For example, some 70 percent of the newspapers in the United States have switched over from lead stereotypes to the tin-free steel plates supplied only by Japan. There have been no restraints in the price increases of these plates. Japan also is the sole supplier of the stainless steel seamless pipe that is an absolute necessity in equipment used under harsh environmental conditions.

One of today's hottest consumer products is the compact disk (CD). Production levels of these disks have exceeded those of

LPs and demand is increasing all the time. Nearly all of the injection-molding machines needed in the production of these disks are made by a single manufacturer located in Nagoya. This firm is working 24 hours a day, but is still unable to meet the demand for its molding machines. The people who make CDs feel that if they had enough molding machines, they would be able to increase their production levels and make a real killing; the Nagoya firm is basically able to set its own prices.

The current appreciation of the yen reveals the route Japan must take in the future. In short, we must develop knowledge — intensive industries characterized by technology — intensive products of the highest added value. This is Japan's only real hope to remain competitive. (Note that production does not have to be limited to high tech products.)

The picture in Japan, however, is not uniform. Even though Japan's manufacturing sector includes firms with the highest productivity levels in the world, productivity in Japan as a whole actually lags behind that of the West. Current run-up of the yen makes some adjustment of these figures, but there is still a gap. The fact is that there are industries in Japan that have very *low* productivity levels; these are pulling down averages for the entire country. These should be thought of as hidden assets in the Japanese company, ripe for improvements. Opening these industries up, one by one, will allow Japan to continue to develop in the future.

The same is true for consumption. Although the Japanese already use money in a very skillful way, they need to think more in terms of the *productivity* of consumption. It is fine for the government to provide companies with grants and assistance, but it is also necessary to consider how that money will be used most effectively. Legend has it that Nero set fire to Rome and enjoyed himself. This was undoubtedly a form of consumption, but anyone could tell at a glance that it was completely selfish, nonproductive consumption. And Rome was destroyed in the bargain.

Domestic investment, in other words, must be carefully targeted for long-term benefits. The electric utilities of Japan have made windfall profits as a result of the current appreciation in the yen. One of these companies spent large sums on new desks, chairs, lockers, and other office equipment for its main office. This was probably better than doing nothing, but it was not a very effective means of returning these profits to its customers. If the money absolutely must be spent, I'd like to see it spent on more forward-looking projects as a means of preparing for an economic winter that is bound to come.

From the perspective of Japanese societal needs, I think it is clear that the concepts of productivity I have discussed earlier need to be introduced into the social realm. Education is one place to start.

One teacher might instruct a class for an hour without the students understanding a word he or she has said. Another teacher might lead the same class, and the students comprehend everything. This points out differences in productivity in education. If students cannot understand something no matter how often they hear it, it may be better to do away with the class. I feel sorry for the children who had to listen without comprehension. Such classes have led to a number of cases of student violence in Japanese schools.

To summarize what I've been trying to say in this chapter, the most important thing we can do is to instill the idea of productivity into all aspects of our lives — whether manufacturing, consumerism, or social systems. This will allow us to avoid the dangers predicted by the NBC broadcast that started this chapter.

One of the major problems of our day is that American industry has lost its competitive power and is getting weaker. These industrial issues have gradually been transformed into political problems, which consequently have become critical barbs aimed at Japan. Those who engender such criticisms are, of course, intelligent people, and even the Japanese are beginning to wonder if they are not correct. But industrial weakness is not America's only problem. Corporate development is the real starting point for solutions to problems of competitiveness and productivity. In the next chapter I address these issues from their logical starting points.

CHAPTER 5

Four Principles of High Productivity

The basis for economic value is manufacturing — making things. But the world functions under a set of harsh competitive principles. You can pat yourself on the back all you want about the outstanding nature of your products, but if someone comes along with a more competitive product, no one is going to buy yours. Product "drawing power" is the real difference in these two cases.

Price is another of the differences between products. To sell, a product must be priced so buyers can afford it; otherwise you've got no product. Delivery time and quality are two other factors to consider when thinking about the drawing power of a product. "Productivity" emerges when you put all these factors together.

On the road to high productivity, there are four milestones: machinery; human beings; control technology; and technological innovation. Productivity can increase only when these milestones are passed and no amount of handwringing can change this rule.

High productivity must be a goal because any firm that has low productivity will most assuredly go bankrupt. The government can jaw at its industries until it is blue in the face and productivity will never increase. For this very reason, countries in Europe are now returning nationalized industries to the private sector.

This is also true in in controlled economies. In Eastern Europe, where the nationalization of industry is a matter of government policy some countries have begun to accept the profit motive. Even the Soviet Union, the bastion of nationalized economies, has already adopted much of the free market concept for agriculture. Finally, China has extended economic freedom further than the Soviet Union and reports one success after another. Trends in labor productivity and the wage cost index for manufacturing are shown in Figures 19 and 20.

The four principles listed above are the key to ensuring increased productivity. What I am about to say may seem very commonplace; these principles, however, are the points of departure for the Japanese and U.S. economies of the future.

Use of Machinery

The first principle, of course, is the use of machinery. Human beings can carry no more than 60 kilos on their backs. Every summer there is much talk about rice prices in Japan. It should be noted that ¥ 20,000 will buy about 60 kilograms, or one bale, of rice, a quantity that would be hard for even a large man to handle.

If this amount of rice were loaded onto a bicycle-drawn cart, however, even an elementary school pupil could pull it along. The job would become easier to do and would be over quickly; productivity would have increased.

One of the reasons for the competitive power of Japanese industry is the fact that it uses machines in so many areas. Nearly 70 percent of the industrial robots of this world are working in Japan.

This large-scale introduction of robots has been important to the mental makeup of the Japanese people as well. In Japanese factories machines are considered friends who help out human beings. It is for this very reason that workers tend to name their machines after popular singers. In the West, though, machines are treated as enemies who have stolen jobs from human beings, and are probably named Frankenstein.

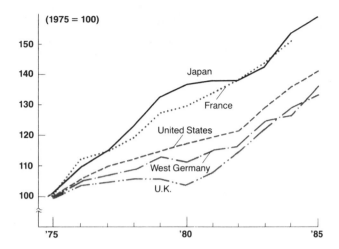

Source: Bank of Japan, *Comparative International Statistics* (1986)

Figure 19. Trends in Labor Productivity in Manufacturing Sector (1975 = 100)

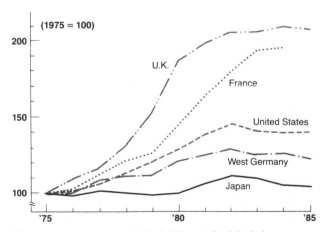

Wage cost index = wage index ÷ labor productivity index
Source: Bank of Japan, *Comparative International Statistics* (1986)

Figure 20. Trends in Wage Cost Index in Manufacturing Sector (1975 = 100)

The Japanese use of robots is increasing with no end in sight. Robots have even been developed recently to prepare sushi. The spread of instant cash machines in banks is another sign of the new popularity of automation. This particular use of auto-mation has advanced to such an intimate point in our lives that it is as if the machines have gotten hold of secret funds that spouses sometimes keep hidden from each other. In many coun-tries, credit cards or checks are the primary forms of payment. Anything that has not been signed by a human being is not con-sidered trustworthy.

But the Japanese seem more flexible and have developed a system that allows the direct debiting of bank accounts for items such as the utility bills; the subscriber does not have to even think about making the payment. It is clear that this in-creasing use of automation has played a major role in develop-ing Japan's high level of productivity.

The Human Being

The second principle concerns the human being. Machines are used by human beings, and they can and will do nothing more or less than a human being has ordered. To increase pro-ductivity, it is important for the people who run the machines to know how to operate them well. Unlike machines, human beings are true individuals with the ability to develop their in-born capabilities. This is a truly marvelous thing.

In Europe and America, on the other hand, scientific control methods are the opposite of those in Japan. There the procliv-ity to treat human beings like machines is commonplace. Thus jobs are subjected to minute analyses, and "standard" jobs are established. People are then supposed to fit into these slots. Time and motion studies, among other methods, have de-veloped from this viewpoint. In Western firms, human beings are merely expected to *respond* to the requirements of their jobs. The feeling is that anything more or less than this could cause problems.

Unfortunately for this philosophy, people are not machines. They are different from day to day. This collision between theory and reality in the West has resulted in their products' being beaten by Japanese products.

Until very recently, Western management methods were being imported without changes into Japan. Suddenly people began to sense their defects. In time these methods were simply dropped and firms rushed to reconstruct their management practices on a different model that includes features such as total worker participation and the creation of small work groups. The result was a new way to facilitate self-awareness.

Identifying and solving human problems such as this are keys to increasing productivity as important as developing employees' capabilities.

Linking Workers and Machines: Control Technology

The third of these principles concerns the question of how best to link the efforts of humans and machines. As noted in Chapter Three, this is known as control technology. Here is a simple example.

Recently Japanese roads have become festooned with traffic lights at virtually every intersection. These lights were installed for safety reasons and are therefore very important; most of the congestion on our roads, however, begins with these traffic lights. When a light finally turns green, the driver may start out, but by the time she gets to the next light, it, too, will be red, and she will have to stop again. This is, I am sure, an experience all too familiar to many readers.

Control technology offered a solution. Engineers observed traffic flow and timed departures from green lights. The lights were then set so that the next light could be made to turn green when the cars from the previous light got close. Then the third light could be set to turn green in a similar fashion, and so on, until nonstop driving was made possible. This systemization of traffic lights can cut the time needed to drive down a street almost in half. A stretch of road that once took 30 minutes could be done in 15, resulting in a 100 percent gain in productivity. I am sure many people would agree that this could reduce the irritation usually associated with driving. In fact, sets of traffic lights of this sort have started to appear recently in Japan's larger cities.

Think about what has really changed in this situation. The road, the traffic light, and the car are all the same. Only one element has changed — the timing of the traffic light in relation to the speed of the car approaching it. The interaction of machine and human has been significantly altered.

Control technology allows us to solve such problems. The excellence of our control technology is one of the reasons Japanese productivity is so high. Quality control is a part of it, as is the kanban method of controlling production.

Earlier I said that a comparative look at factories in the West and Japan will show that there are no major differences in facilities or products. In spite of these similarities there are major differences between Western and Japanese goods in terms of quality, costs, and productivity. The reason for these differences is a puzzle that cannot be fully comprehended with just a brief look at the factories. The solution lies in control technology.

In the United States, for example, General Motors and Toyota have established a joint production facility called New United Motor Manufacturing Incorporated (NUMMI) in Fremont, California. Under GM's earlier management, low productivity at this plant and terrible worker-management relations had ultimately closed this plant. One of the stipulations in reopening the plant was that NUMMI employ members of the United Auto Workers. This was fulfilled and members of the UAW are now making cars in the plant.

Despite its earlier history, productivity at this plant is clearly higher than similar plants, even though there has not been any large-scale introduction of robots. Quality, moreover, equals that of imported Japanese cars; it is much higher than that of comparable U.S.-made cars. There is concrete evidence to back up these facts. In contrast, a highly modern plant that GM

The NUMMI automobile factory

built at great expense in Hamtramck, Michigan, has performed poorly. Nine months after its completion it was producing at only 70 percent of planned capacity.

An article entitled "Detroit Stumbles on Its Way to the Future" in *Business Week* (June 16, 1986) compared the two plants and noted that the key to NUMMI's high productivity is the Toyota management style, "which emphasizes participative management, lean layers of middle management, and decision-making pushed as close as possible to the assembly line" (p. 104).

The Fremont factory is an experimental plant for GM. A separate building near the plant houses GM representatives who come to the plant each day and faithfully record what is going on there, with particular attention paid to what the Japanese are doing. Unfortunately, it is not that simple; management is not something that can be understood merely from daily observation of what transpires on the factory floor.

The heart of the matter revolves around jobs that are tied together organically: What types of ideas tie people together? What appropriate actions should be taken under situations that arise only occasionally? There are many similar inquiries.

Daily, routine activities do not tell the whole productivity story. A Toyota official told the observers that they would never obtain a full understanding of the situation no matter how carefully they watched.

Under the current UAW contract, each worker's job is described in detail; complete agreement has been reached as to exactly what each job should constitute. While each union member can be asked to do every one of the tasks contained in his job description, he or she can neither be asked to do any other job, nor to perform at a level higher than that of the job description. Wages are determined under the "pay-for-work" rule.

The overall system appears on first glance to be rational. But many irrational things can happen in a factory every day. Some are virtually impossible to foresee.

In a real factory, it is not all that easy to allocate jobs. Parts don't come in when they are supposed to, tools break unexpectedly, and machines stop running. Problems are discovered and design changes ordered. People don't show up for work, defects start appearing for which the cause is not known, and so on. Every day is a new battle. If procedures for all of these problems are not carefully thought out, the factory can grind to a complete halt.

How does this work at the NUMMI plant? Part of the answer involves individual responsibility. Any worker on any line in the plant is empowered to stop the line if he or she discovers anything wrong. This avoids the potentially serious effects of allowing a defective part to flow on to the next step in the process.

Steeped as they are in the customs of American management, it is hard for workers to fully accept these tenets. Under the NUMMI system it is no longer enough for them to simply perform their own jobs. If they do not cooperate to restart the line, not a single car will get made. At this point they begin to move naturally toward something with which they have absolutely no experience — direct participation in reforms that improve the level of factory operation.

In the manufacturing business, a single mistake can cause the whole system to crash. Even if no one makes a mistake, problems can still arise from causes that no one ever anticipated.

This is exactly the same situation faced by a ship's crew. If the ship sails into trouble and begins to sink, you can be sure that it will go straight down if everyone adopts the attitude that all they have to do is perform their own job. Everyone must pool their strength and cooperate to save the ship. In dire situations, even the most beautifully written job description won't do a bit of good.

This approach is really just a matter of common sense. Once they got used to it, the NUMMI employees began doing marvelous work. These workers came to realize the total emptiness of "scientific control" methods previously in effect. Under that regime, logic would suggest that job analysis on a person-by-person basis and the subsequent skillful assignment and combination of these jobs should lead to the most efficient.

But human beings are not machines. They want to feel pride and accomplishment in the work they do. This sense of craftsmanship develops only if there is a certain amount of freedom within a job. This loose way of doing things probably appears irrational to the Western mind; it is a fact that this concept of "gray" has helped Japanese factories to succeed.

For example, the standard type of conveyor used for electronic assembly lines in Japan is called the "free flow" line. In this system the flow of parts is stopped while one person is doing his or her job and resumed when the person is finished. It is a way of adjusting the machines to the rhythm of human beings. Anyone can tell that this is a better way of doing things without even stopping to think about it, but by building on these common sense factors one at a time the Japanese factory was able to succeed.

This is one of the secrets of Japanese management. The Japanese method has restored a sense of craftsmanship in the workplace. This has made people more enthusiastic about their jobs and instills in them a sense of pride that is missing in many American plants.

These phenomena, of course, have not gone unnoticed in the United States. Japan's industrial successes have spurred the popularity of a rash of "theories concerning Japanese management." As far as I was concerned, though, only the effects of the theories were analyzed; the essence of the topic was missed. This is because any attempt to fully understand what is essentially a "soft," human process solely within the "hard," conceptual framework of American management practices cannot be fully successful. Attempts to absorb these practices without really knowing what has gone into them are doomed to failure.

For example, consider just-in-time manufacturing, famous the world over. Before the development of this system, factories would purchase quantities of materials and parts, store them in a stockroom, and withdraw them for use when needed. Of course, that means creating storage space for parts and materials that are not used for some time. A large capital outlay is required, and the idle parts must be insured.

No matter how you look at it, this is stupid, so manufacturers began purchasing only what they needed "just-in-time" to produce a given product. This cut storage losses and related costs to nothing. Anyone can understand the idea behind just-in-time manufacturing. Even so, the system cannot be implemented until a number of preconditions have been met.

First, parts and materials acquired from suppliers must be free of any defects. The line will be stopped dead if even a single defective part gets to it.

Second, setup times must be short. In just-in-time manufacturing, a variety of different items are produced on a single line; the system cannot deal successfully with this variety if setup times are too long. If possible, setup time should be no more than 10 minutes; otherwise, idle time increases and productivity falls off drastically.

Finally, a company must have a product with proven sales attraction and enough orders to keep the production line running at full capacity. Anything less makes the just-in-time system a bunch of nonsense.

There are, moreover, a number of examples in the United States where kanban control methods failed because they were implemented by firms that did not fully understand the requirements for just-in-time production. For every success story about a company that raised productivity and saved hundreds of thousands of dollars a month, there are other factories where the introduction of kanban methods without a thorough understanding of just-in-time threw the manufacturing processes into utter turmoil and shattered productivity. After listening to all these conflicting stories, some Americans have concluded that just-in-time manufacturing is mere fantasy, made up by the Japanese to hoodwink American management.

In truth, the simpler a given principle, the more necessary it is to optimize conditions in the surrounding environment; if this is not skillfully done, the system will not function properly. Again, the secret is control technology. Only that allows the manager to deal one at a time with each set of data from the work area.

I have now dealt with three of the principles of higher productivity: use of machinery, development of human potential, and the control technology that links the first two together organically. The final principle is technological innovation.

Technological Innovation

At an OECD conference held in Paris in 1985 I made the following remarks: "A Japanese television station recently sent a team of investigative reporters out to discover how counterfeit wristwatches bearing famous European brand names are being made in Hong Kong and other newly industrialized countries. One of the things they did on this program was to make a careful comparison between the real watch and the counterfeits. In point of fact, the fake was very well made, and on the outside it was impossible to tell the difference between it and an original. But there were differences.

"The first of these was the way the second hand moved. The original had a mechanical movement, and the second hand thus moved ahead smoothly, in very small increments. But the fakes were quartz, so the second hand moved more slowly, in increments of one second at a time. If you knew that, you could tell immediately which one was the fake. But there were other differences as well: Because the original was mechanical, it lost a minimum of 10 seconds a day. But the fakes, being quartz, lost less than two seconds a day. Also, they cost only one-tenth of the original.

"These distinctions show how frightening technological innovation can be."

This appeared to shock my listeners quite a bit and for a while, no one asked me any questions. Japan has succeeded in technological innovation and can say things like this — consider how awful we would feel if we were on the receiving end of such practices. It's clear, I think, that whatever course Japan takes in the future, it will have to be centered around technological innovation.

There are many misconceptions concerning technological innovation, especially as it occurs in Japan. One of these is that technology is somehow the property of specialists who ensure that it cannot be easily understood from the outside. A second is that there is no need for these specialists to explain technology to people in understandable terms.

The next stage in the friction between the United States and Japan, I am convinced, will center around technology. If Japan reacts unskillfully to this technological friction, matters could turn out as they did in my example concerning watches. There are ample reasons to fear that such problems will not be solved by simple economic effort or negotiations. Although America has often been bested by Japan in the trade arena, most Americans believe that they will win in any technological dispute; many people in Japan share that feeling.

Things are not so simple in the real world. To understand the role of technological innovation in increasing productivity we must know exactly what technology *is*. Chapter Six takes up this question in detail.

CHAPTER 6

Establishing a Technological Country: Japan's Choices

As Japan has become more and more successful, we have seen more and louder debates — both in Japan and abroad — on the nature of the technology that has supported that success.

There are those who maintain, half in jealousy, that virtually all of Japan's technology was originally found in the West, and that all Japan has done is to use it somewhat skillfully. Similarly, there are those who maintain that Japan has made very few contributions to science, which forms the basis for technology. These people aver that unless Japan engages in more fundamental research, it will be doomed to be forever pegged with the pejorative term "a nation of imitators." No, others say, the problem lies in the creative abilities of the Japanese people — Japan must be transformed into a nation that encourages more creativity. In this ongoing debate there is no shortage of opinions.

Distinguishing Technology and Science

Like the trade friction between the U.S. and Japan, this is a question that demands study and policies from a variety of different angles. To avoid the danger of meaningless debate, I would like to start with a more fundamental set of questions: What is "technology"? How does it differ from and relate to "science"? Is technological progress possible without new developments in science? Though I may appear to be starting from a rather elevated position, I do not intend to introduce a host of difficult philosophical issues. All I wish to discuss are matters of common sense, things we can easily understand by simply looking around us.

The human race has discerned a large degree of regularity in the world of natural phenomena. From this scientists have derived certain physical laws. In putting these laws to practical use, humanity has provided itself with useful products. We continue today to acquire more and more new knowledge from nature, and thereby expand our understanding of natural law.

The purpose of technology, however, is to make things; technology often employs principles that are hardly new. Furthermore, failures can occur no matter how good the specific underlying ideas. Technology does not fully develop until ideas have been transformed into products and the products sold. In other words, technology succeeds or fails in terms of its implementation.

I was graduated from a university during World War II. At that time I bowed to my youth and spent a lot of time reading books. After the war I noted the appearance of a rash of new materials, machining methods, and products, all under the banner of "technological innovation"; I always had the feeling that I had heard about them somewhere before.

The path to successful implementation can be a long one. Someone may use the principles determined by someone else and attempt to make something for some use or other. At that stage, however, it is only an idea, an idea that will die if it cannot be transformed into a thing. Let's say it does so in a certain case. Years later, perhaps, someone else will take up the problem and this time make something that appears to work for a particular purpose. In turn, this too may die from lack of materials, high costs, or other reasons. Even if materials are available and new machining methods have been devised, the product may be sold as a prototype only to die because no one will buy it.

Finally, after several years have passed and circumstances have changed, there may be a demand for the product. Production technology will be established, and an industrialist will set sights on its manufacture. To the world at large the product may then seem to have sprung out of nowhere to quick acceptance in the marketplace.

The above sequence describes the way I felt about the technical innovations that took place in the 1950s. My feelings have not changed. Biotechnology and electronics have received a good deal of notice recently. Some say they are completely new. Even in these fields, however, you can find many principles and products that can be traced back for more than 20 years. They have appeared in the light of day only after passing through a lengthy development and testing period.

For many, the glittering achievements of recent technological innovation appear as a sudden flowering of a garden of new plants. These innovations, in many cases, had their start before the 1970s, when people proclaimed that technology had reached its limits.

Even today, we are sowing the seeds of the technology of the future. New discoveries alone, however, are not the seeds for new technology. There are any number of technological innovations appearing in splendor today, due to advances in science or machining technology, that are based on principles laid out a century ago.

The quartz watch is an example of a technological revolution brought about by converting a century-old idea into a thing. Pierre Curie first discovered that the piezoelectrical qualities of a quartz crystal allow it to generate electricity. This was in 1883.

The vacuum tube was created and quartz crystal clocks were developed. The earliest models, unfortunately, were nearly the size of supply cabinets. They were certainly accurate, but a little difficult to put on one's wrist!

Thirty years later, advances in semiconductor technology gave birth to the large-scale integrated circuit (LSI). The LSI made a quartz wristwatch feasible and the first models appeared in Japan. Cheap versions of these watches are going for under ¥ 500. They lose less than two seconds a day.

Another example is the development of LCDs. Molecules of normal liquids are aligned totally without order. Liquid crystal, on the other hand, is aligned with regularity in only one direction. This discovery was made quite some time ago. In the 1960s RCA first announced an idea that involved the use of liquid crystal for the display of letters. The liquid crystal of that time, however, was slow; the displays were not clear, and the idea proved unusable.

In the early 1970s, the pocket calculator boom started in Japan. The first of these calculators used green fluorescent displays; this tended to gobble electricity. The "battery-life wars" began — as soon as one company advertised that its batteries would last for 1,500 hours, another would claim a life of 1,800 hours, and so on. As long as you are using fluorescent display tubes, however, you are bound by their 2,000 hour limit. In contrast, liquid crystal consumes scarcely any electricity. All of the calculator companies leaped on the liquid crystal bandwagon. They started using it not only for pocket calculators, but for clocks and watches that demanded low power consumption.

Even a liquid crystal television screen has been produced as an extension of these technological developments. A liquid crystal television set can be thin and flat. In the future it is likely that TV sets will hang on walls like paintings. This would be a true revolution in television. It should be remembered, of course, that the principles behind liquid crystals were first discovered more than 100 years ago.

There's another lesson to learn here. Developments in quartz crystal and liquid crystal have been greatly facilitated by the advances made in semiconductor technology. In short, their compatibility with other technologies is good and it is truly a matter of good timing.

These examples show that science and technology inhabit essentially different realms and have different goals. People often speak of technology as flowering on a bed prepared by science. This implies that the two are actually the same thing — and from there a number of misconceptions have arisen.

There is no question that it will be technology that will support the Japanese and American economies in the years to come. It is therefore important, in my opinion, for people who are not specialists in technology to have a correct understanding of its significance. With this in mind, I'd like to relate the tale that follows.

In March 1986 I got a telephone call from the U.S. Embassy — "Mr. Bloch is looking for you," I was told. It turned out to be Eric Bloch, who had been in charge of semiconductor development at IBM and was now the head of the Board of Directors of the National Science Foundation. What does he want with me, I wondered. I found out that an important invitation had arrived for me from E. D. Maynard, Jr., head of the VHSIC (very high speed integrated circuit) project then being conducted by the Department of Defense.

As I remarked earlier, the hollowing of American industries such as semiconductors and machine tools — industries vital to the national defense — has resulted in these firms losing out in competition with foreign producers. The Department of Defense increasingly relied on foreign companies in these areas. This, of course, presented a major risk to America's national security. Eric Bloch was a member of a policy committee that had been formed to study the problem, and the committee wanted to hear my opinions.

The revitalization of U.S. industry is one of my most pressing concerns. Since I was extremely busy at the time I decided to jam a trip into a tight schedule. I would catch a plane on Thursday, appear at the meeting in Washington on Friday, and come back to Japan on a plane the next day.

This was not to be a public hearing, and so it would be possible to speak freely. The participants were limited to the top levels of American industry and included engineers; we were able to have a full and free exchange of opinions.

The meeting centered on the question of why American technology, believed to have the lead in international developments, was having so many problems and what could be done to get it back on track.

Since the infamous National Semiconductor fiasco of February 1984, the American semiconductor industry has faced other embarrassments. In 1985, three firms were accused of supplying defective parts to the military and were heavily fined by the Department of Defense.

The recent explosion of the Challenger Space Shuttle is still reverberating in our ears. The successive list of management problems that were revealed in that context clearly represents the difficulty in integrating critical technologies that have grown increasingly more specialized in nature. I tried at that Washington meeting to present the American participants with a consciousness of how technology should be understood.

Understanding Technology

Americans in the semiconductor business display somewhat ambivalent feelings concerning Japan. The source, I am sure, is their knowledge that America is the country that started it all. Furthermore, many new inventions were made by Americans. Somewhere along the line, American products have been pushed aside by their Japanese competitors. Japan has now cornered some 90 percent of the world market for 256 K RAM chips. Americans are as irritated as a person who has just been fooled by a magician. They have noticed that many of the semiconductors being used in U.S. military equipment were made in Japan. Suddenly, they have to rely on foreign products for critical defense equipment.

At the Washington meeting, I began my attempt at explaining what technology really is with a discussion of the rotary engine. This engine was originally invented by a German. It uses only a small triangular rotor and needs neither a complex spindle nor a crankshaft. Although it is small in size, its RPMs can be increased so that it can produce high horsepower. Engine makers the world over jumped to take advantage of this wonderful idea. It proved remarkably difficult to work out in reality; engine burnouts were the biggest problem.

The piston engine has a wonderful component called the piston ring that effectively prevents engine burnout. It efficiently circulates the lubricant in addition to keeping the cylinder airtight. Rotary engines have no piston rings. Many manufacturers were burned in trying to deal with rotary engine burnouts; one by one they fell by the wayside. One famous Japanese specialist in aircraft engines even went so far as to say that the Japanese had been tricked by the Germans and that there was no way that this engine would ever succeed. But finally, one firm in the world — Toyo Kogyo (now Mazda) — succeeded in making a commercial version of this engine despite the fact that even in Germany, where the engine was invented, there had been no success in commercial development!

The rotary engine was most certainly a German idea, but I don't believe we can get away by merely saying that the Japanese added "just a little bit of an improvement" to it. After all, Mazda was the only corporation in the world that was able to develop the technology needed to produce this engine. No matter how wonderful an idea might be, it cannot be classed as "technology" until it has been converted into a useful product.

This story contains several excellent lessons in the essence of technology. The first of these is that no matter how great something might be in principle, principle alone will not convert it to a thing. Technology is a method used for the purpose of making things; if you cannot make anything, you don't have it. The idea for the rotary engine may have been born in Germany, but the technology that permitted the actual construction of the rotary engine was born in Japan.

The second lesson is that it takes a firm commitment from top management to actually develop technology. Though other firms had fallen by the wayside one after another, it was President Matsuda's firm commitment that allowed the development of the rotary engine.

The third lesson is that the success of this engine required the discovery of a new material for the rotor seal and lubricant. In piston engines, the spindle ring provides the cylinder seal and lubrication. In the rotary engine, though, sealed areas would freeze up; this is where the other companies failed in their development attempts. Mazda succeeded by developing a new material to solve this problem. This proved to be the key factor in determining the success of the rotary engine. Technological advances do not necessarily have to come from the most current discoveries, and neither did Mazda's new material.

The fourth lesson lies in the importance of luck. At the very instant Mazda's engine was perfected and began selling in the United States, the world was hit by a fuel crisis. The rotary engine suffers from bad fuel economy, and sales fell off drastically. The company nearly went broke but persevered. Had Mazda stopped production of the engine at that point, it never would have become a reality. All of Mazda's efforts did finally pay off. The engine is used in its RX-7 sports model. This car became a runaway best seller and remains very popular.

In order to ensure success, all the technological ingredients necessary to allow the underlying principle to be converted into a product must be present and ready to roll. Top management must be committed to the project. But luck, or as I use the term, timing, must also be present.

If the timing is off, even a technology destined for glorious success can end in excruciating failure. The history of technology contains countless examples of outstanding technological seeds that were never brought to fruition because of poor timing. The successes of Japanese technology today are flowers blooming in the graveyard of the countless failures of the past. They are not cut flowers from a garden labeled "imitations of things foreign."

The semiconductor industry presents the same lessons. For example, a material known as gallium arsenide allowed one to overcome the speed barrier once present in silicon semiconductors. People had known about this ultra-high-speed material for some time, but it was Japan that was able to actually apply this material on a chip and provide it to the market.

The gallium chip is now an indispensable part of the high-performance radar and high-speed computers used in weapons.

It is virtually the sole property of Japan because only Japanese manufacturers have proven capable of manufacturing these chips at the high degree of reliability demanded for weapons. The United States has no choice but to buy them, no matter how unpalatable this may be. This illustrates better than anything else the technological differences between the United States and Japan.

The Americans make a big deal out of who invented something or other. They say very little about who made it, something I find rather strange. In science, victory certainly goes to the person with foresight. In technology, however, you have nothing until a manufacturer makes an actual product, actually uses it, and it performs up to snuff. A television is not a television, for example, until someone flips the "on" switch and an image appears on the screen in the right way. If a television suddenly started smoking when you turned on the switch you might want to call it a jack-in-the-box; it certainly could not be called a television. That is what I stressed at the meeting in Washington.

These simple principles are surprisingly difficult to understand with just a brief glance. Identical principles are being used to manufacture similar products in all countries. Even so, there are clear differences in areas such as quality, cost, delivery time, and the like. The reason for these differences is remarkably difficult to comprehend if all you do is look.

When Japan first succeeded with transistor radios, there were people who attributed this success to the manual dexterity of the Japanese people. Today, modern industry has few products that rely on the manual dexterity of the people who make them.

There are also people who say that Japan's successes come from the fact that it has plenty of funds available for capital investment. Expensive new equipment in itself is not the linchpin of successful production, as the earlier comparison of NUMMI and the GM-Hamtramck plants pointed out. Yet another explanation is that most of the outstanding engineering graduates in the United States go into the defense industries, and few join the private sector. For this reason, it is maintained, consumer products made in the United States will never be competitive with Japanese products. I have very little faith in any of these explanations for the poor showing of the United States economy.

If the equipment, the people, and the products are all basically the same, where do the differences come from? At the Washington meeting I said:

"The major, fundamental problem is in the different ways that American and Japanese engineers think."

Winning the Battle Against Gray

Whenever you argue with American engineers, their analysis inevitably takes on a "binary" perspective. By this I mean that things are either black or white, yes or no — or they break things into a number of different categories and try to force everything into one of them. In manufacturing, you will find aspects that are neither black nor white but gray. Reality is a constant struggle with these gray areas; trying to force them to become either black or white is ignoring reality.

I respond to binary thinking with the following example from real life. At an American-run plant in Singapore, I once opened up a defective semiconductor to try to find the reason for the failure. I found a damp patch on the inside that was clearly either a drop of sweat or spit from the person who had worked on the product. Because of that drop, the aluminum elements had begun to corrode.

The plant manager was in Japan at the time and I showed him a photograph we had taken of the semiconductor. I asked him to take more protective measure with the clothing worn by the workers at his plant. He ignored my request and said that because this particular semiconductor had a protective film on its surface exposure to moderate amounts of contaminants in the air would not affect it. Even though I protested that we had a photograph that clearly showed what the problem was, he refused to accept what I said.

The fact is, even if a product has a protective film, that is no assurance that it is perfect. In this case, a Japanese would think that even if what this manager had said were true, the true effect of contaminants cannot really be ascertained, and that's why we consider them to be essentially bad. We do our best to totally eliminate them. The American manager reasoned that because the semiconductors were coated with a protective film, contamination was eliminated as a reason for defects!

This type of binary thinking can be a problem; no matter how impressive your logic might seem, if the product is defective, so is the logic. A logic without contradictions had taken shape in the head of this American. Not only would it be unthinkable for him to take even one step forward from this world, but the world itself could exist nowhere except in his own head. In the real world, it is rather rare to find things broken up clearly into rights and lefts. To do so ourselves is simply a matter of convenience that allows us to construct a neat theory.

On this point, I had more to say to the committee in Washington. Creative production is a daily struggle against error; the target you have planned may be wrong; there may be a mistake in the design; the materials you have ordered may not be exactly right; machine operations can't always be trusted. Only by dealing with such errors one at a time can you maintain consistently high standards.

The average American engineer, however, has the extremely naive notion that all one must do is follow standards and blueprints to ensure that everything will be okay. I then told the committee the following story to show the errors inherent in that approach.

A Japanese rubber firm was buying belts from an American company and selling them. Some of the belts broke and a claim was filed. The Japanese firm called an American engineer to investigate the matter. The engineer said that the belts all met standards and were not defective; if any had broken, the fault must have rested with the user. He stubbornly refused to listen to the Japanese side of the issue. This intransigent attitude continued a long time. Finally, a year after the event, the Japanese firm received its customer's okay and subjected the belts to a thorough examination. The problem was found and corrected in a mere two weeks.

After quoting this example, I said something even worse: "There are many deities in Japan — in fact, we speak of the 'myriads of gods' in our literature. If your value system won't admit the idea that the world contains a large number of different truths, rather than one single truth, then I don't think you'll ever succeed in creating anything."

In other words, to win the production race one must first consider the existence of the gray areas, then figure out how to deal with them. The Japanese have never really let these gray areas bother them very much; they don't resist their existence. This attitude has been manifested in the patterns of religious belief based on an amalgamation of Buddhist and Shinto beliefs. The Christian missionaries saw this as heretical, but the Japanese attitude is just the opposite. However well this two-dimensional framework might serve logical argument, it is a positive menace to the creativity that is derived from establishing a flexible relationship with reality.

The Japanese believe that all employees must pull together in the struggle with the gray areas. Because the area is gray, it cannot be dissected with our conventional technological sense and theories and we may not immediately know what parts of it are optimal and what harmful. There should be some point along the continuum with which we can work comfortably. Japanese employees work together to discover where that point lies. This is done primarily through two mechanisms: the quality control circle, and the avoidance of overspecialization in worker job descriptions. The first of these is a Japanese invention that has only recently been tried in the West.

Overspecialization, on the other hand, is a hallmark of Western management. In the West, management has developed towards a system of specialists. At first glance, this appears rational, but these specialists are expected to become all-knowing, all-powerful gods in their own area of expertise; no mistakes are permitted. Make a mistake and you are out on your ear the next day. When someone makes a mistake, he or she will rationalize it away somehow in his head and avoid taking the responsibility for it. Again, production is a struggle against reality. You might be able to fool a human being; you cannot fool a product.

This, I stressed to the Washington committee, is the major reason that the American semiconductor manufacturers are being beaten out by the Japanese. I said, "There are currently a number of American-run firms with semiconductor plants in Japan, and all of these plants are making outstanding products. Texas Instruments of Japan has a plant in Oita that last year won the Deming Prize, the top award for quality control in Japan. I've been there myself, and it's certainly an outstanding place. But the facilities at that plant are no different than those at other Texas Instruments plants throughout the world, the products they are making are identical, and of course the designs are the same, too. Even so, there is a major difference between this plant and the others. Why? It's because the employees are Japanese, people who believe that every problem is everyone's problem, people who are engaged in a constant struggle against gray areas and error."

The key that will unlock the door to leading edge technology is right here. In the next section I'd like to expand on this theme.

Surprising Sources of Leading Edge Technology

After the first energy crisis, Japan's economy entered a low-growth period with annual growth in the GNP between 4 and 5 percent. This caused some people to make statements like the following: "During the period of high growth, everyone had to run, or we wouldn't have made it. But now we've entered a period of low growth, so why don't we slow down the pace to a walk for a while?"

This proposal may look plausible to some, but as far as I'm concerned, it is completely wrong. Let us look at growth rates by industry. There are several industries in Japan whose growth rates are low, or even negative. The aluminum and soda industries are in big trouble because of high energy costs in their production. The Japanese steel industry has the most outstanding technology in the world, but it is growing at a rate of only between 1 and 2 percent.

Despite the fact that we have stagnating industries, the overall GNP is growing. There must be some industries with growth rates higher than the GNP, or the figures would not jive. If you take a closer look, several will apply. The electrical equipment industry is recording double-digit growth. The automobile industry is also propelling the Japanese economy down the road of growth.

The overall composition of Japanese industry as well as the internal makeup of specific sectors is changing rapidly. I said that electrical equipment is doing well, but that is not true for the entire industry. Lighting is off because of the drop in demand for homes. Computers and video equipment have grown rapidly. The internal composition of the electrical equipment industry is changing at a very fast clip.

I believe that there are four forces that combined are bringing about these changes in the internal composition of industry in Japan. In many ways they apply as well to the United Statses. The first of these is technological innovation (introduced in Chapter Five). The second is the changing social environment. In 1985 the health industry generated revenues in excess of ¥ 1 trillion, and the aging of the Japanese population was the main reason for this increase. The third force derives from changes in the laws and systems of the country. Last year changes in the laws governing electrical communications changed demand patterns for telephones. A whole new line of businesses appeared thereafter. The last of these four forces is the wave of internationalism that has swept over Japan. All these are apparent in Figure 21.

Of all these forces, however, technical innovation will be the most important source of new business. The ceramic industry is a prime example. It is said that the recent strengthening of the yen has put that industry in a terrible position; even under these circumstances there are ceramics manufacturers doing quite well. They have either done well in the "new ceramics" fields or are making "novelty" products. The latter, in particular, are of high added-value and can be made in no other country in the world. Some of these products, such as ceramic packages for

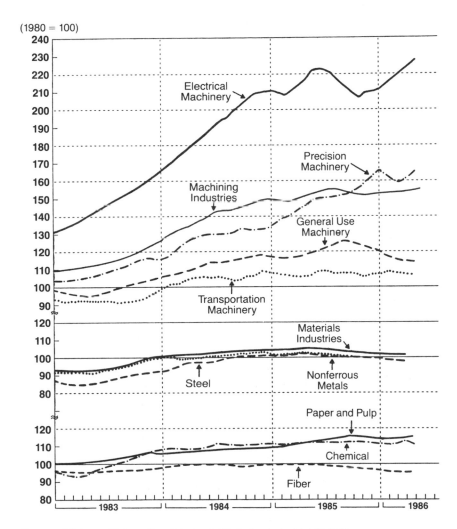

(1980 = 100)

Source: Ministry of International Trade and Industry, *Industrial Statistics*
1. Machining industries include general use machinery, electrical machinery, transportation machinery, and precision machinery. Materials industries include steel, nonferrous metals, chemicals, papers and pulp, and fiber.
2. Each index reflects average three-month trends for seasonally adjusted index.

Figure 21. Japanese Production Trends by Industry

semiconductors, sell for ¥ 10 or more per gram. Automobiles go for about ¥ 2 per gram and even videocassette recorders for only about ¥ 7 per gram.

Technological innovation can also add value to more prosaic products. The other day I visited a company that makes the lead frames used for the leader lines for semiconductor assembly. These can be made by stamping a thin plate. When I asked how much they sold for, I was told that one gram was about ¥ 3. Consider how much hard work goes into the making of a complex automobile, which sells for a mere ¥ 2 per gram. The reason is that these lead frames are incredibly precise — accuracy up to .001 millimeter is guaranteed! They may look simple, but a great deal of expertise lies behind their manufacture.

Technological innovation is often seen in even everyday consumer products in Japan. The zirconia-based ceramic scissors and other ceramic cutting tools developed in Japan are good examples. Ceramic scissors are able to cut synthetic fibers that metallic scissors cannot. They can cut through thin titanium

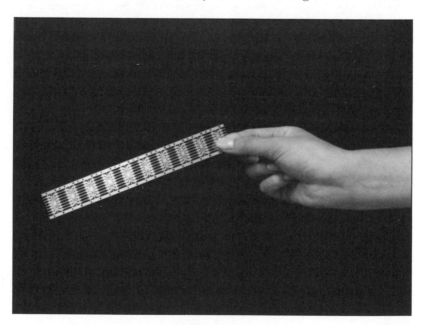

Lead frame for semiconductor assembly

plates as if through paper, something that could not be done by hand before without the aid of a special press.

Most of these products are being made by small and medium-sized businesses. These firms do not necessarily employ engineers with a lot of high-level credentials. In fact, it is often common middle-aged men and women who work for small and medium-sized companies that crank out these wonderful, high-level technological products. I know they sometimes have a hard time controlling the contempt they feel for those square-shouldered Western engineers who often belittle their efforts. More than anyone else, it is such people who have succeeded in the struggle with the gray areas.

In 1985, I attended a symposium on leading edge technology held in Toulouse, France, where I had this to say: "Westerners seem to have some sort of misconception of leading edge technology. Because you think that it's something terribly difficult, you tend to use it in difficult applications, like missiles or space development. But because 'technology' somehow has been linked to 'difficult,' it seldom actually gets used in consumer product form."

The Japanese, on the other hand, are willing to search everywhere for a place to use leading edge technology once it has been developed. Carbon fiber, which is lighter than aluminum and stronger than steel, is a good case in point. This fiber was a major discovery made by a person at the industrial testing center in Hyogo Prefecture. It took some time to develop its production; once that was done, it was first used in golf clubs, then in fishing rods. In such consumer applications, there is little problem if a few defects are uncovered. After the kinks were worked out and mass production techniques were perfected, the fiber was used in the production of airplanes.

The application of "shape memory" alloy is another good example. The shape of this alloy changes with temperature. Japanese manufacturers first used it in coffee makers (to allow hot water to flow over the coffee grounds when it reaches a certain temperature), then in air conditioners (to automatically adjust the direction of air flow). Finally, they began to use it in brassieres. When the brassiere is folded and placed in a dresser drawer, the material folds right up; when put on, it expands to its original shape. This is an incredible invention and Japan is far and away the world production leader today.

An earlier and more fundamental development, the LSI, also has an instructive history. Originally, LSIs were developed for the American space exploration program. Production runs for this use, or for military use, were in the five-figure range at the highest. The chips cost ¥30,000 or ¥40,000 per unit. It was a Japanese calculator manufacturer who first noticed the possibilities of the LSI. Their use allowed the company to make calculators small enough to put in your pocket. Used in calculators, the monthly LSI production levels soared into the millions. The price dropped into the range of several hundred yen per chip. After calculators, LSIs further developed for use in quartz watches and finally microcomputers.

The United States Department of Defense has never understood the technological development of the LSI and its adaptation for use by the common people. To them, LSIs had significance only as they were used by the military. Quality and performance standards were thus seen as somehow separate from those of products made for the daily lives of the common people; the military bought LSIs made with especially high specifications. This is stupid, since there are no significant differences in performance requirements between products meant for the military and products meant for civilians. The LSIs are the same.

Ironically, products that are made in the awesome quantities demanded for consumer mass production may well be more reliable. Equipment investments further are repaid rapidly and companies can install the most up-to-date machinery. In comparison, the standards for military products are slow to be updated and refined. Production technology for consumer products now exceeds that of military products.

In the February 4, 1985, issue of the trade magazine *Electronics Weekly*, three articles by engineers appeared in succession. The theme of all three was that military standards are responsible for fettering the progress of the U.S. semiconductor industry. Despite the fact that microcomputers are available, the military continues to use old-fashioned ICs. Manufacturers are therefore forced to go on building equipment that would be more appropriately placed in museums. In these authors' opinion, this is an "insane" situation.

Contrary to American thinking, technological innovation does not preclude imitation. People in the United States talk about the "NIH syndrome." The term "NIH" stands for "not invented here," and the idea behind it is that there is no point in producing a product that was not invented in your own country. In other words, Westerners tend to be more interested in congratulating themselves for their own foresight than they are in selling things. Sales come later.

Japan, though, is exactly the opposite. Imitation is encouraged as another way of fostering innovation. The other day an American congressman asked me a question: "If a company comes up with a hit product in Japan, other companies are making exactly the same thing within three months. Doesn't this indicate that there is some sort of collusion on the part of the manufacturers?"

I told him, "No, you have to remember that the engineers employed by the Japanese firms are friends from college, or know each other well from attending various meetings together. So if one company comes up with a best seller, this is terrible. The president of all the competing companies will call in his engineers and let them have it — 'That was a friend of yours who made that product. Why can't you do it?' After being raked over the coals by his boss, the engineer will get mad — 'If that turkey can do something like this, then I can too!' And that's why all the other companies suddenly come up with the same basic thing."

This imitative adaptation of technology is, in my opinion, a uniquely Japanese phenomenon. Consider the VCR. The Americans developed one that sold for ¥20 million. The Japanese made one for use in the home for a few hundred thousand yen. Tell American engineers to lower the cost of something to one-hundredth of its former price and most would think it was impossible. But Japanese managers have dreams. They imagine the demand there would be if a VCR could be put in every home! So they work as if possessed to develop products for home use. With VCRs, this took incredible amounts of seed money and a lot of time, but finally resulted in a tangible product.

This use of technology for the masses will continue to be important to Japan in the future, and there is no question that Japan will lead the world in this area. But this cannot take place unless we have a firm base of research and development behind it. In the next few pages, I'd like you to consider this last point.

Research and Development Fills the Gap: The "Moon Dust" Compound

In recent months, the amount of money being invested in research and development in Japan has begun to increase rapidly. An unpublished report of January 1986 by the U.S. Department of Commerce begins its section on Japan with a look at how many new research facilities with staffs of 500 or more have been opened in Japan over the past five years. The upsurge shown in these figures gives us a good idea of the pace at which Japan is apt to be traveling in the future.

In addition, R&D as a percentage of total costs has also increased. In the past, for example, the medical supply field commonly believed that one's R&D investment should be 10 percent or more of total sales; these days even electronics firms have gotten to the 10 percent level. If you consider the impact on any manufacturer who lays out 10 percent of operating income on R&D you might get an inkling of the load this kind of investment places on companies.

Heavy R&D investment, however, allows companies to announce the development of new technology almost daily. Product mix has also changed. Products that were the mainstay of company sales five years ago now account for only 20 to 50 percent, at most, of total sales. The extent to which new technology is vital to the average firm could not be made more clear.

If this type of thing keeps up, smaller firms may doubtless begin to feel out of the race. Can anyone besides the largest firms afford to spend R&D money at that level? "We smaller firms just can't manage it," might well become a common thought.

Some areas of research, of course, are impossible to pursue without spending large amounts of money. Neverthless, wondrous inventions are also possible without spending a cent. I'd like for you to know why, since the answer is another crucial element in the successful application of technological innovation.

After much work, the American Apollo program finally managed to send a human being to the moon in 1969. This took an incredible amount of money to accomplish. The moon landing taught us a great deal more about the moon and a massive number of reports were published on the topic. Thanks to these reports, a Japanese made $50,000. The following narrative relates how it happened.

Before Apollo, there had been a great deal of speculation concerning the condition of the surface of the moon. There were many theories. One said that the surface of the moon was buried in a deep layer of dust from that falling constantly from space. Some of this space dust even falls toward earth, but it disappears as it hits the atmosphere.

The surface of the moon, however, is a vacuum. Thus, the theory held, the dust would have built up there over the 4.5 billion years since the moon was formed. Given the resultant dust buildup, some people feared that the rocket could sink down and never be able to leave again.

When the Apollo rocket actually arrived on the moon, it turned out that this was not the case. One of the reports published on the topic speculated that the dust flying through space is moving at an incredible speed, fast enough to give it an electrical charge. When it comes into contact with the grounding surface of the moon, the effect is similar to what would happen if a soft rice cake were splattered against a wall. It would adhere tightly to the surface. Hence, there would be no deep layers of dust on the moon.

A Japanese person read this article. He was immediately struck with an idea. Wouldn't it be possible to make a new alloy in this fashion? To do so he created a vacuum and introduced metallic particles, then applied voltage and caused these particles to swirl about. This operation created a new alloy, and he patented the idea and sold it in the United States for $50,000.

There is a very important lesson here. A large amount of money was indeed necessary for the discovery of this phenomenon. It was, of course, NASA's money. Tens of thousands of people throughout the world must have seen this paper; only this single person thought of using this process to create a compound. This person was able to come up with his idea because he knew the complexities involved in the making of compounds. In other words, the first important thing is to know the problem.

Earlier I noted that when new technology is developed in the West people want to use it in difficult applications; the reason is that Western engineers are familiar with these difficult topics. In Japan, however, product development is done from a "market-in" approach and engineers are best acquainted with the problems in this area. Their familiarity with the activities of daily life provide the stimulation for the development of many innovative products that meet the needs and desires of consumers.

The Bottleneck Approach to Technology:
The MIG 25 and the Alpha-7000

Japanese corporations provide many opportunities for their engineers to come into direct contact with users. When times are hard and products stop selling, some companies even mobilize factory workers as well as technological development engineers to go out and sell. This practice allows firms to circumvent "bottlenecks" in technology and markets and find technological solutions.

When the automaker Mazda was at a low point, it reassigned more than 10,000 people, including engineers and factory workers, as sales personnel. This invaluable experience helped these employees to go back to their normal jobs and design an extremely successful car that brought Mazda back to life. This was an incredible accomplishment and one difficult for foreigners to comprehend. In the United States, companies simply lay off everyone they don't need at such times. The firm's outstanding engineers will then be snatched up by competitors faster than rats abandoning a sinking ship.

If all firms were like this, it would make little difference. When competitors like the Japanese — who think totally differently — come along, the differences are extremely obvious. I call this the "bottleneck approach to technology."

As long as there are engineers in Japan who continue to circumvent bottlenecks in consumer-oriented technology in this way, this approach to technological development will continue to provide the country with one success after another.

The bottleneck approach to technology is not, of course, the sole property of Japan. The Soviet Union, surprisingly enough, also has practiced this method. The story is as follows.

In the 1970s, the MIG 25 was faster than any airplane in the West, including those of the United States. When it made its first appearance in the skies over the Mideast, American Phantom fighters were unable to catch it; when they finally did get a radar reading of its speed, the result greatly shocked the U.S. Air Force.

Fortunately, a Soviet defector landed a MIG 25 in Japan in September 1976 and the United States was able to examine an intact specimen. The air force heaved a great sigh of relief when the plane turned out to be something like a rocket, with a fuselage of stretched sheet steel and a heavy nose. They finally realized that although it was capable of great bursts of speed, its weight seriously limited its range. It was not, however, until they actually examined a MIG 25 that the American experts realized how the Soviets had managed to convey the illusion of technological superiority.

The MIG 25 was a last-ditch response to U.S. Mach 3 bombers. In matching U.S. speeds, the temperature of the plane would increase to the point where duralumin could not be used. Titanium was, at that time, unobtainable. Steel sheeting managed to provide the strength for high speeds, but the extra weight meant that flying time had to be sacrificed in order to provide the jet with the speed it needed to pursue the American bombers.

The United States was not aware of this design solution. The Soviet Union decided to send some of the MIG 25s across the noses of U.S. Phantoms in the Mideast to provide a speed demonstration. America was completely taken in by this maneuver. If the Soviet pilot had not defected, the nightmares faced by the U.S. Air Force over this airplane would have continued indefinitely.

This episode says something interesting about technology. The Soviet Union must be seen as being somewhat behind the U.S. and Western European countries in ultrasonic airplane technology. From the twin perspectives of how technology is used, and which side has the military superiority, however, it is difficult to say which side is superior. Performance is, of course, important in the development of military equipment, but the ability to psychologically threaten the other side is also an important factor. The Soviet Union discovered a way to overcome their weakness — the bottleneck in their technology — with technology already in place. That solution was the MIG 25 and its steel skin.

The bottleneck approach to technology can also be applied to consumer products, of course. One of the latest hit products in Japan is the Minolta Alpha-7000, an automatic-focus SLR camera. When it was first released, one camera specialist noted that it contained no new technology; its excellence lay in the way it implemented features that were already available. "This camera should sell," the specialist said. At the time the Alpha-7000 camera was first sold, SLRs were seen primarily as the province of real camera fanatics. SLR demand had slackened and production had dropped. The new Minolta camera was a big success. It breathed new life into single lens cameras and a boom to the stagnating camera industry. In short, this camera was successful in overcoming the bottlenecks that had been present prior to its production.

Leading edge technology is in the news these days, but technological innovation does not have to be beyond the reach of your company. I'd like to stress this to more than just Japanese companies. Everyone could profit from these lessons, whether the Western industrial nations, the newly industrialized countries of Asia, or the developing countries. You will only block progress if you complain about your lack of one technological thing or another or try to blame your own failures on someone else.

In the first six chapters of this book I have outlined the causes of trade friction between the United States and Japan and presented the analytical and management principles that will allow American managers to rectify their problems of competitiveness and productivity. Keeping all these lessons in mind, I would like to present my own proposal for the course we should take in the future.

CHAPTER 7

Sharing the Wealth by Exporting Japanese Expertise

In the summer of 1985, a study meeting in the city of Fuji-Yoshida was sponsored by the Japanese Federation of Employers' Associations (Nikkeiren). Someone asked me, "What do you think about Japanese agriculture?"

I answered very bluntly: "What about Japanese agriculture? Given the proper methods, it should be a wonderful thing indeed. Japan has better natural conditions than any nation on earth. It doesn't get hot and it doesn't get cold, and since we're surrounded by the ocean, our climatic changes are very stable. In addition, we get plenty of rain that we need for agriculture. Our soil is also good — just inviting agriculture. If you go to Korea, for example, the soil is so old that there are hardly any trees on the mountains anymore. Recently, I went to Spain, and when I entered the country from France, and saw all that expanse of red dirt, I thought that this must be the only type of soil for vineyards.

"Any country that's this blessed by natural conditions should be able to triumph in competition with foreign agriculture. If we can't, that's clearly the fault of man and not the fault of nature."

I am sure you will already understand what I was driving at. As I've mentioned before, the people living on the Japanese archipelago have the highest educational standards in the world and are very peaceful and hard-working people. For these very reasons, more and more of our industries are taking their places in the top ranks of world manufacturing. Even industries and countries now at the lower levels of the international pecking order should be able to develop their competitive power.

In reality there are many barriers to economic progress in Japan. One of the most formidable of these is the current aging of the population; a distinct tapering off at the bottom is now occurring. Age distribution tables of public and private organizations have until now been shaped like a pyramid, a shape made possible by the even age distribution of the population. But as you can see from Figure 22, this has changed; the pyramid now resembles a pipe with a crimped base. We must start making policies now that will be needed to deal with this problem in the future.

Secondly, Japan will not be allowed to be the only winner at the Mah-Jongg table forever. The developed and the developing nations must all find a route that will allow them to coexist in prosperity. One of the policies that will facilitate such co-

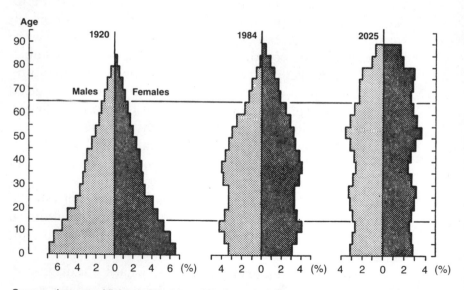

Source: Japanese Ministry of Health and Welfare, Institute of Population Problems

Figure 22. The Japanese Population Pyramid 1920-2025

prosperity is for the developing nations to follow Japan's example described in Chapter Three: become suppliers of high value-added manmade resources.

Transferring Japanese Management Techniques

A third action is the most important. This will be to transfer management practices and expertise from Japan to other parts of the world. These lessons have been learned through years of experimentation and hard work and are now ready for export.

After World War II, Japan became an experimental field station, so to speak, for the world's most modern management practices. After Japan's defeat in the war, it wished more than anything to catch up with the advanced nations; to this end management began the direct importation of any and all industrial methods from abroad; some methods were worthwhile, and others evinced absolutely no grasp of reality. In those days Japan wasn't very choosy, however, and took in everything that came along. By now, only the worthwhile techniques have survived.

Massive losses were sustained during these experiments, but the country ultimately succeeded. It is now time for Japan to share with the rest of the world the knowledge and expertise that we have built up over the years.

These methods even include some techniques that were abandoned in their country of origin, but eventually implemented successfully in Japan. For example, the concept of quality control has been extremely successful in Japan. In the United States — the country that gave birth to the statistical methods that underlie quality control today — it has not been practiced on nearly as broad a scale as in Japan. Dr. W. E. Deming, who provided so much direction to Japan in the early days of quality control, was not known very widely in the United States until the 1980s. He was credited on the NBC program (discussed earlier in Chapter Four) as the person responsible for the success of quality control in Japan. His name became known throughout the United States overnight. He is now an extremely busy octogenarian!

I have many opportunities to view factories around the world. Very often I leave with the feeling that a factory would succeed if it would only implement one program or another.

Recently, in fact, I visited a semiconductor factory that is supported heavily by the government of Taiwan. My wife was accompanying me. At the door to the clean room, where dirt or dust of any sort is absolutely abhorrent, we were given white anti-dust robes to wear. My wife was carrying a handbag and asked what she should do with it. I told her to put it down somewhere, since she shouldn't bring it into the clean room. But the plant manager said: "Oh, that's okay, just come on in with it."

I was absolutely shocked. Handbags and other personal effects are veritable storehouses of contaminants. In Japan no one would ever dream of taking one into a clean room. A Japanese firm that bought out an American factory in Silicon Valley told me it took two years to get the workers to take off their shoes before they entered a clean room. After my experience in Taiwan I understood what they were talking about.

The expertise gained from simple cases like this, where just one look is all it takes to see what's wrong, can be utilized in high-level analyses and solutions. The result will be an improvement in the technological standards of all nations and production of high-quality goods at low costs. This will conserve natural resources as well as make a direct contribution to the enrichment of society.

The necessity for clean production areas is not limited to the semiconductor industry. At a Japanese factory producing tape decks for cars, assembly line workers are required to change their work gloves twice a day. This may appear wasteful, but the nap from the gloves had been found to cause product failures. Stopping failures before they have a chance to happen makes a lot of sense in terms of both paring costs and preserving the company's reputation among consumers. Of course, the used gloves are not simply thrown away after they've been changed; they are recycled to other, less critical, jobs.

These examples point to the fact that the high reputation enjoyed by Japanese products would not be possible if companies failed to thoroughly investigate all possible causes of product failures and to develop policies to counter them.

When I relate these tales to foreign factory managers, they usually give me a look that says, "There's no need to get that uptight about it!" When I show them photographs of actual breakdowns and supply figures to show the effects of production improvements, they usually come to understand what I mean.

A Theory of "Gray Technology": Cartesian Error

As I have said many times, the making of things is a constant battle with the gray areas. I have touched upon many aspects of this ideas throughout the book, but I think it is worthwhile to reiterate its importance here in the development of both the technology and the management philosophies that ensure high productivity and error-free production.

Recently, I discussed this concept with Gene Gregory, Professor of International Business at Sophia University in Japan. In the course of our conversation, he made the following insight. "The binary view of the world that divides everything up into blacks and whites started with Descartes, and was certainly indispensable as a starting point for the solving of scientific problems. But even though this has been a useful tool in our quest for unknown worlds, I have the feeling that it hasn't been so very useful as a principle of leadership in fields of corporate endeavor, such as manufacturing and sales."

As you can see, his opinions and mine are identical. Descartes' ideas say it is possible to skillfully extract regularity from what appears, at first glance, to be chaotic natural phenomena, and having illuminated these aspects of nature, it is possible to derive generalized rules from them. Cartesian ideas have certainly functioned effectively as the fundamental guiding principle of scientific inquiry. In fact, some believe that the reason that China never developed the discipline of natural science is due to the fact that it never applied Descartes' ideas.

While critical to science, this philosophy proves difficult to apply to corporate activity. To make or sell a specific product, for example, requires that you consider an almost infinite number of factors, any one of which can affect whether or not you will achieve your goals. They include product performance; price; competing products; distribution; the overall economy; and many others too numerous to count. The task is not just to construct a theoretical model that will take all of these factors into account. The Cartesian solution, so to speak, is to extract those factors expected to have the most significant effects on operation and construct a predictive model based on these alone.

Unfortunately, in the real world of business there is no guarantee that the one factor that was overlooked will not have a significant effect on the outcome. The error that results from these overlooked factors falls into the "gray areas" — its cause is not identified by a limited model.

The traditional Western approach is ineffective in eliminating error in many cases. It may be useful in common academic research to construct a set of artificial and very narrow conditions that will reduce the possibility of error by as great a degree as possible. This "controlled cultivation" is permissible if it advances our knowledge, but often it leads to dead ends — generalizations that do not hold true in the real world. In the world of corporate management — and even in the worlds of politics and economics — those gray areas have generally been acknowledged, and attempts made to find successful methods to deal with them.

Professor Chie Nakane of Tokyo University has said that Japanese society is characterized by a lack of formal regulations. It is my belief, nevertheless, that the reason for the success of the Japanese economy is that the Japanese have constructed a set of methods for dealing positively with error and gray areas. These methods have been constructed and brought to a fine polish by the Japanese themselves. They have then been brought to bear in the battle against the variable conditions of the real world. In sumo wrestling this is referred to as "the art of fighting without a set style."

In this process of development, Cartesian thinking has been modified to take gray areas into account. There are people who say that just-in-time production was already present outside Japan before the Toyota production system was developed and, similarly, that quality control was originally imported from the United States. Japan reworked these ideas through extensive trial and error to make them more suitable for the factory. Just as they transformed science into usable technology, they refined these concepts so that they actually serve a purpose. In short, both just-in-time and QC were products of Descartes' way of thinking; it was the Japanese who made them flexible or "gray" enough to be used in the factory as tools for the craftsman.

The potential for error is everywhere. In the manufacturing plant the human being is a clean slate; error is the clothing he or she wears. Automated machinery does not always work properly when we flip the switch. In sales, we are always working with human beings; determining whether they will buy or not involves a potential for error that is considerably larger than that faced by a person who works in a manufacturing plant. Error lurks at every step of the way.

Corporations try not to approach this difficult state of affairs as a gamble, but rely on their employees' experience and intuition to get them through. Some companies have very high rates of success; others meet with one difficulty after another, run up a debt, and eventually collapse. In the final analysis, there are skillful ways of doing things and not-so-skillful ways. The determining factor is inevitably not Cartesian logic, but whether or not management has developed a system that recognizes and deals with gray concepts to keep the company healthy. This is a fundamental of business administration, diplomacy, and any other endeavor where we must deal with error.

As I noted toward the beginning of this chapter, when you stop to consider the situation Japan truly has been an experimental country. A lack of preconceived management notions has enabled the country to test virtually every hypothesis and proposal made by others throughout the world. One can figure out almost immediately whether they're the real thing or just imitations of reality just by putting them into practice. The Japanese will happily throw out anything that doesn't work or change it thoroughly if it appears to have some promise. This is the attitude of the craftsman and is completely different from that of scholars relying on established knowledge; such craftsmen hold on to nothing simply for its own sake.

A Final Wish

At this point, I would like to express a fond wish of mine to the reader. This wish is that the Japanese evaluate, analyze, and, most important, share their management expertise with the rest of the world. The results should be reduced trade friction and greater prosperity.

This is no time to be stingy with our knowledge or to worry about some sort of boomerang effect. Japan is not the originator of this knowledge, but the refiner. We have absolutely no need to cling to what we have to save face; we are not the founders of a religion. It is more important that what we have accumulated can be put to use to make the world a richer and fuller place for all people.

In a case like this it may be appropriate to receive small recompense. I am talking here about the export of Japanese wisdom and expertise, not products, and hope that this is not going to wind up as some idiotic exercise in trade friction.

Cultural pride and friction may interfere with this plan. There may be those who will refuse our call because of their pride or mistrust. We should make videotapes to give to such people, since it is only human nature for them to want a peek at the goods, so to speak.

By sharing our experiential data, we will mitigate the worst aspects of trade friction. The object is to bring the manufacturing performance of other countries up to Japan's level.

The basic essence of this trade friction is that our customers are starting to sink, and they want to pull us down with them. Again, the semiconductor industry is a perfect microcosm. Great steps were taken in the 1970s to lower the price of semiconductors and improve technology in that field. We have seen with our own eyes, however, the 1986 cooperative agreement between Japan and the United States put the brakes on that progress. The 1 megabit memory chip is upon us and prices are going to be forced up to incredible heights. That is likely to drive U.S. computer manufacturers back to outmoded 256 K technology. What will the people of the next generation think of this, do you suppose?

The 1986 semiconductor trade agreement does not prohibit Japanese manufacturers from selling chips cheaply within Japan. It won't be long before we can export products that use these chips at prices cheaper than the chips themselves. This is competitiveness at its most basic level and it is clear that it will start trouble. The agreement is a result of the binary thinking that must express things as either "fair" or "unfair." It was completely asinine.

It would be much better if we made some effort to move forward. In other words, we should give the world a hand, instead of sinking with other declining manufacturing economies. I simply cannot understand why this type of proposal almost never comes out of Japan. People must think that if they offer something like this other countries will get mad at them, so they keep quiet.

Keeping quiet is like trying to force the hands of a clock to go backwards. There are sensible listeners in the United States and Europe who want progress on these issues. If anything, in fact, they far outnumber those who sit around splitting hairs and carping day and night. The former are widespread in both Europe and the United States. France, in particular, is working very enthusiastically. In traveling through England, Austria, Holland, Spain, and other countries, I have seen considerable evidence of a willingness to improve productivity and competitiveness through management and not political methods.

In Chapter Six I described how I was invited by the U.S. Department of Defense to address these questions. I spoke very frankly and touched on things that normally would cause a sharp response. Even so, my listeners were open to what I said, and the results were extremely good. Further, two of the six proposals that I made for American industry were implemented within a month.

All of the world economies today share a common fate, bound together by economic ties that cannot be cut no matter how hard we might try. Japan, fortunately, has been successful with its own industries and now must strive as hard as it can to enlarge the world's economic pie. To this end, I personally am now acting as chairman of the government's International Technological Strategy Research Group. We are considering ways that we can export Japan's expertise to the rest of the world.

The "concept of gray" is critical to this expansion, because it is the best way of implementing the technological innovation that underlies the world economy. Change occurs rapidly in technology these days. If we take our foot off the accelerator even a little, as we did in the case of semiconductors, we will surely slide backwards. Japan must keep its foot on the gas, but at the same time extend a hand to other countries to help them accelerate. This is the one and only way we can keep the economic clock moving in the right direction.

To ensure success, we must take on Descartes, so to speak, and assert the value of understanding the "concept of gray." Among the intellectual methods used to deal with the gray areas of the world is "groupism." It is critical to problem solving. There are people who deny this by noting that creativity is only really lodged in the individual. But it has been said for a long time in Japan that "three people can have the wisdom of Monju" (the Bodhisattva of wisdom). It is natural enough that foreign countries have emphasized the individual; after all, they were the originators of individualism. There are countless examples, however, of major inventions resulting from a relaxed conversation.

The individual-group dichotomy can be argued forever. Let's instead set our sights on what are essentially more important things. What is important to remember is that the "concept of gray" is a practical "here-and-now-ism" that always proceeds on the basis of reality.

Reality is the major touchstone of both economics and technology. No matter how well thought out, a bit of logic is only an empty exercise if things don't work out the way it predicts they will. In the old days, when it was hard to obtain new information, there was little doubt in the effectiveness of the principles advanced by those in leadership positions. Now we are in an age where we can get information whenever and wherever we need it; it's time to change our approach to solving productivity and competitiveness problems.

The cybernetics of Norbert Wiener is one such new approach. Before Wiener, people had thought that making minute changes in *theories* (even those that have not worked successfully in the past) will help them work more successfully. There are still people who think this way. Wiener responded to this by stressing that if you perform all actions on the basis of feedback from reality, rather than from theories, you will be more successful.

Here is an example of cybernetic thinking in action. In the old days, people tended to think that if a cannon shot did not destroy its target, all that was needed was to adjust the aim. Today's computer-controlled missiles chase moving targets and every shot can be a direct hit. In this situation, it is not necessary to aim precisely. Just point your cannon in the right direction and fire! Errors that pop up will be rapidly corrected in response to on-the-spot data. This is what I mean when I speak of dealing with the gray areas.

If you trace Japan's progress since World War II, you will see that this is exactly what has been done. In the postwar period the Japanese did not have any particularly farsighted ideas. They just did what they had to do every day; in the process, the country's GNP rose to 10 percent of the GNP of the entire world. If Japan had clung to esoteric philosophical principles, this would never have happened.

Japan can continue to prosper using these simple principles. Furthermore, so can the world. A certain flexibility of spirit is required. If the Japanese can keep this flexibility, appreciation of the yen can be overcome without disaster. If the United States and Europe become more flexible on trade friction and management, they too will benefit.

In this situation, there is not much point in attempting to predict what direction the world economy will take. Who can tell the direction the waves will carry us? What's important is that we develop the ability to firmly grasp and effectively deal with the reality before us. The flexible operation of a management system based on typical Japanese conceptions is an important step down the road toward economic success. It will not work for us to place a set of vague principles in front of us and proceed on the idea that we must merely choose one of two alternatives. We need instead to gut it out together.

The novelist Soseki Natsume said it best: "The world will never get anywhere unless we occasionally read Shakespeare backwards, and dare to say that what we've seen there is worthless."

About the Author

Hajime Karatsu was born in Hyogo Prefecture, Japan, in 1919. He graduated in 1942 from the Imperial University of Tokyo with a degree in electrical engineering. In 1948 he entered Nippon Telegraph and Telephone Public Corporation. He joined Matsushita Communication Industrial Co., Ltd., in 1961, advancing to director in 1971 and managing director in 1978. In 1984 he became a technical advisor to Matsushita Electric Industrial Co., Ltd., and in 1986 he joined the faculty of Tokai University's R & D Institute. Mr. Karatsu received a Deming Prize in 1981 for his work in quality control and was honored by the Ministry of Education in 1984 for his distinguished contributions to the development of industrial education in Japan. This is his first book to be published in English.

Index

Other Fine Books from Productivity Press

Productivity Press publishes and distributes materials on productivity, quality improvement, and employee involvement for business and industry, academia, and the general market. Many of our products are direct source materials from Japan that have been translated into English for the first time and are available exclusively from Productivity. We also sponsor conferences, seminars, in-house training programs, and industrial study missions to Japan. Send for our free catalog.

Tough Words for American Industry

by Hajime Karatsu

Let's stop "Japan bashing" and take a good close look at ourselves instead! Here is an analysis of the friction caused by recent trade imbalances between the United States and Japan — from the Japanese point of view. Written by one of Japan's most respected economic spokesmen, this insightful and provocative book outlines the problems and the solutions that Karatsu thinks the U.S. should consider as we face the critical challenge of our economic future. For anyone involved in manufacturing or interested in economic policy, this is a rare opportunity to find out what "the other side" thinks.
ISBN 0-915299-25-9 / $24.95

Today and Tomorrow

by Henry Ford

Originally published in 1926, this autobiography by the world's most famous automaker reveals the thinking that changed industry forever. Long out of print, the book has been largely forgotten. Yet Ford's ideas have never stopped having an impact; in fact, today they are re-emerging to revitalize American industry. Here is the man who doubled wages, cut the price of a car in half, and produced over 2 million units a year. Time has not diminished the progressiveness of his business philosophy, or his profound influence on worldwide industry. You will be enlightened by what you read, and intrigued by the words of this colorful and remarkable man.
ISBN 0-915299-36-4 / $24.95

Productivity Press, Dept. BK, P.O. Box 3007, Cambridge, MA 02140 (617) 497-5146

Manager Revolution!
A Guide to Survival in Today's Changing Workplace
by Yoshio Hatakeyama

An extraordinary blueprint for effective management, here is a step-by-step guide to improving your skills, both in everyday performance and in long-term planning. *Manager Revolution!* explores in detail the basics of the Japanese success story and proves that it is readily transferable to other settings. Written by the president of the Japan Management Association, and a bestseller in Japan, here is a survival kit for beginning and seasoned managers alike. Each chapter includes case studies, checklists, and self-tests.
ISBN 0-915299-10-0 / $24.95

Toyota Production System
Beyond Large-Scale Production
by Taiichi Ohno

If you have ever wondered what the famous Toyota Production System is all about, this is the book for you. It's the original — the first information ever published in Japan on the Toyota Production System (also known as Just-In-Time manufacturing). Not only that, it was written by the man who created JIT for Toyota. In this book, recently translated into English, you will learn about the origins and development of the system, as well as its underlying philosophy. Any company aspiring to be a world class manufacturer must use the concepts developed by Mr. Ohno. This classic is the place to start.
ISBN 0-915299-14-3 / $39.95

Inside Corporate Japan
The Art of Fumble-Free Management
by David J. Lu

A major advance in the effort to increase our understanding of Japan, this book shows *why* Japanese businesses are run as they are — and how American companies can put this knowledge to good use. Lu has spent many years in Japan, personally knows many top leaders in industry and government, and writes with a unique bicultural perspective. His very readable book is full of anecdotes, case studies, interviews, and careful scholarship. He paints a well-rounded picture of the underlying dynamics of successful Japanese companies. *Inside Corporate Japan* is a timely and invaluable addition to your library.
ISBN 0-915299-16-X / $24.95

Productivity Press, Dept. BK, P.O. Box 3007, Cambridge, MA 02140 (617) 497-5146

The Sayings of Shigeo Shingo
Key Strategies for Plant Improvement
by Shigeo Shingo, translated by Andrew P. Dillon

A recent issue of *Quality Digest* claims that Shigeo Shingo "is an unquestioned genius — the Thomas Edison of Japan." The author, a world-renowned expert on manufacturing, "offers new ways to discover the root causes of manufacturing problems. These discoveries can set in motion the chain of cause and effect, leading to greatly increased productivity." By means of hundreds of fascinating real-life examples, Shingo describes many simple ways to identify, analyze, and solve problems in the workplace. This is an accessible, readable, and helpful book for anyone who wants to improve productivity.
ISBN 0-915299-15-1 / $36.95

Productivity Press, Dept. BK, P.O. Box 3007, Cambridge, MA 02140 (617) 497-5146

BOOKS AVAILABLE FROM PRODUCTIVITY PRESS

Christopher, William F. **Productivity Measurement Handbook**
ISBN 0-915299-05-4 / 1983 / 680 pages / looseleaf / $137.95

Fukuda, Ryuji. **Managerial Engineering: Techniques for Improving Quality and Productivity in the Workplace**
ISBN 0-915299-09-7 / 1984 / 206 pages / hardcover / $34.95

Hatakeyama, Yoshio. **Manager Revolution! A Guide to Survival in Today's Changing Workplace**
ISBN 0-915299-10-0 / 1984 / 198 pages / hardcover / $24.95

Japan Management Association and Constance E. Dyer. **Canon Production System: Creative Involvement of the Total Workforce**
ISBN 0-915299-06-2 / 1987 / 251 pages / hardcover / $36.95

Japan Management Association. **Kanban and Just-In-Time at Toyota: Management Begins at the Workplace,** *translated by David J. Lu*
ISBN 0-915299-08-9 / 1986 / 186 pages / hardcover / $29.95

Lu, David J. **Inside Corporate Japan: The Art of Fumble-Free Management**
ISBN 0-915299-16-X / 1987 / 278 pages / hardcover / $24.95

Ohno, Taiichi. **Toyota Production System: Beyond Large-Scale Production**
ISBN 0-915299-14-3 / 1988 / 176 pages / hardcover / $39.95

Ohno, Taiichi. **Workplace Management**
ISBN 0-915299-19-4 / 1988 / 176 pages / hardcover / $34.95

Shingo, Shigeo. **A Revolution in Manufacturing: The SMED System,** *translated by Andrew P. Dillon*
ISBN 0-915299-03-8 / 1985 / 383 pages / hardcover / $65.00

Shingo, Shigeo. **Zero Quality Control: Source Inspection and the Poka-Yoke System,** *translated by Andrew P. Dillon*
ISBN 0-915299-07-0 / 1986 / 328 pages / hardcover / $65.00

Shingo, Shigeo. **The Sayings of Shigeo Shingo: Key Strategies for Plant Improvement,** *translated by Andrew P. Dillon*
ISBN 0-915299-15-1 / 1987 / 207 pages / hardcover / $36.95

AUDIO-VISUAL PROGRAMS

Shingo, Shigeo. **The SMED System,** *translated by Andrew P. Dillon*
ISBN 0-915299-11-9 / slides / $749.00
ISBN 0-915299-27-5 / video / $749.00

Shingo, Shigeo. **The Poka-Yoke System,** *translated by Andrew P. Dillon*
ISBN 0-915299-13-5 / slides / $749.00
ISBN 0-915299-28-3 / video / $749.00

Productivity Press, Dept. BK, P.O. Box 3007, Cambridge, MA 02140 (617) 497-5146

SPRING/SUMMER BOOKS FROM PRODUCTIVITY PRESS

Ford, Henry. **Today and Tomorrow** (originally published 1926)
ISBN 0-915299-36-4 / May 1988 / $24.95

Karatsu, Hajime. **TQC Wisdom of Japan: Managing for Total Quality Control**
ISBN 0-915299-18-6 / June 1988 / $34.95

Karatsu, Hajime. **Tough Words for American Industry**
ISBN 0-915299-25-9 / May 1988 / $24.95

Mizuno, Shigeru (ed.) **Management for Quality Improvement: The 7 New QC Tools**
ISBN 0-915299-29-1 / May 1988 / $59.95

Ohno, Taiichi and Setsuo Mito. **Just-In-Time for Today and Tomorrow: A Total Management System**
ISBN 0-915299-20-8 / August 1988 / $34.95 (tent.)

Shingo, Shigeo. **Non-Stock Production: The Shingo System for Continuous Improvement**
ISBN 0-915299-30-5 / June 1988 / $75.00

Shinohara, Isao (ed.) **New Production System: JIT Crossing Industry Boundaries**
ISBN 0-915299-21-6 / May 1988 / $34.95

TO ORDER: Write, phone or fax Productivity Press, Dept. BK, P.O. Box 3007, Cambridge, MA 02140, phone 617/497-5146, fax 617/868-3524. Send check or charge to your credit card (American Express, Visa, MasterCard accepted). Include street address for UPS delivery.

U.S. ORDERS: Add $3 shipping for first book, $1 each additional. CT residents add 7.5% and MA residents 5% sales tax. Add $5 for each AV.

FOREIGN ORDERS: Payment must be made in U.S. dollars. For Canadian orders, add $8 shipping for first book, $2 each additional. Orders to other countries are on a proforma basis; please indicate shipping method desired.

NOTE: Prices subject to change without notice.

Productivity Press, Dept. BK, P.O. Box 3007, Cambridge, MA 02140 (617) 497-5146